Methods
for
Skin Absorption

Editors

Barbara W. Kemppainen, Ph.D.
Assistant Professor
Departments of Physiology and Pharmacology
College of Veterinary Medicine
Auburn University
Auburn, Alabama

William G. Reifenrath, Ph.D.
Research Chemist
Division of Cutaneous Hazards
Letterman Army Institute of Research
Presidio of San Francisco, California

CRC Press
Boca Raton   Ann Arbor   Boston

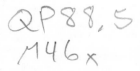

Library of Congress Cataloging-in-Publication Data

Methods for skin absorption / editors, Barbara W. Kemppainen and William G. Reifenrath
p.   cm.
Includes bibliographies and index.
ISBN 0-8493-4651-7
1. Skin—Absorption. 2. Methods
SB781.L68 1990 632'.11—dc19   88-30252

Direct all inquiries to CRC Press, Inc., 2000 Corporate Blvd., N.W., Boca Raton, Florida, 33431.

© 1990 by CRC Press, Inc.

International Standard Book Number 0-8493-4651-7

Library of Congress Card Number 88-30252
Printed in the United States

PREFACE

Anyone who has embarked on a new area of research is probably familiar with feelings of intimidation as one begins to review the literature. Skin penetration has been studied since the 19th century, however in the last 10 years, the area has grown tremendously. Several books and reviews have been published covering general aspects of percutaneous absorption. While these resources serve as valuable guides, one is forced to consult the original literature when designing experimental protocols. The investigator is faced with a bewildering assortment of techniques and apparatus for making skin absorption measurements; he or she may acquire the hopeless feeling that the entire specialized literature needs to be mastered. Often it is extremely helpful to visit a laboratory where similar research is being done or to call investigators with specific questions. In this manner, one can learn at least one approach to a problem and often important details that have escaped review or even publication. This book was conceived in the spirit of the laboratory visit. Its chapters provide rough paths through an appparent jungle of methods. As the investigator travels these paths, he can smooth some of the bumps, enlarge the narrow places, and even create new trails.

Since the chapters for this book were written concurrently, some duplication was inevitable, particularly for closely related topics. We have chosen to retain the overlap to provide alternate views on a given subject and also to help preserve the continuity of individual chapters. The final chapter is set forth to play devil's advocate. That is, to question the existing methodology in the hope of stimulating development of new techniques or refinenment of the old methodologies to overcome limitations.

We were fortunate to have the cooperation of a group of authors active in their fields of specialization who still took the time to contribute their chapters for the book. We are indebted to them for the high caliber of their work, the patience they displayed in dealing with our suggestions, and their willingness to meet deadlines which minimized the delay between writing and publication.

Barbara W. Kemppainen
William G. Reifenrath

THE EDITORS

Barbara W. Kemppainen, Ph.D., is Assistant Professor of Toxicology at Auburn University College of Veterinary Medicine at Auburn, Alabama. Dr. Kemppainen obtained her B.S. degree from Ashland College, Ashland, Ohio in 1976, M.S. degree from Ohio State University, Columbus, Ohio in 1978, and Ph.D. degree from the University of Georgia, Athens, Georgia in 1982. She was awarded a U.S. Department of Agriculture Research Associateship from 1982 to 1986, and served as Assistant Professor at Auburn Universtiy School of Pharmacy from 1986 to 1988. It was in 1988 that she assumed her present position.

Dr. Kemppainen is a member of the American Academy of Veterinary and Comparative Toxicology, American Society for Pharmacology and Experimental Therapeutics, Society of Toxicology, and Southeastern Regional Chapter of the Society of Toxicology (SRCSOT). She served as Councilor (1986) and President (1989) of SRCSOT. She has been the recipient of two research contracts from the U.S. Army Medical Research and Development Command.

Dr. Kemppainen has presented 15 invited lectures, published 15 research papers and two book chapters. Her current research and teaching interests include skin and mucosal absorption of xenobiotics and effects of mycotoxins on animals.

William G. Reifenrath, Ph.D., is a research chemist in the Division of Cutaneous Hazards, Letterman Army Institute of Research, Presidio of San Francisco. Dr. Reifenrath obtained his training at the University of Nebraska, Lincoln, receiving his B.S. degree in Chemistry in 1969. He was a National Science Foundation Trainee from 1969 to 1974 and received his Ph.D. in Pharmaceutical Science in 1975. He was a postdoctoral fellow at the University of the Pacific School of Pharmacy from 1975 to 1976 and served as Captain, Medical Service Corps, from 1976 to 1980 at Letterman Army Institute of Research. In 1980, he assumed his present position.

Dr. Reifenrath is a member of the American Chemical Society, American Association of Pharmaceutical Scientists, and the American Society for Testing and Materials. He has served as an advisor to various Army, Air Force, and other federal agencies dealing with skin protection and decontamination and chemical protective clothing. He has been a National Research Council Fellowship Advisor since 1987 and is a member of Rho Chi and Phi Beta Kappa honorary societies.

Dr. Reifenrath is the author or co-author of more than 30 scientific publications. His current major interests are mechanisms of skin absorption and improved models for assessing occupational exposure to hazardous chemicals.

CONTRIBUTORS

Charan R. Behl, Ph.D.
Research Investigator
Pharmaceutical Research and
   Development
Hoffman-LaRoche, Inc.
Nutley, New Jersey

Kenneth E. Black, M.D.
Dermatologist
Department of Medicine
USA MEDDAC
Fort Ord, California

Robert L. Bronaugh. Ph.D.
Supervisory Research Pharmacologist
Division of Toxicology
U.S. Food and Drug Administration
Washington, D.C.

Hing Char, Ph.D.
Senior Scientist
Pharmaceutical Research and
   Development
Hoffman-LaRoche, Inc.
Nutley, New Jersey

Stephen W. Frantz, Ph.D.
Group Leader
Department of Pharmacokinetics
Bushy Run Research Center
Export, Pennsylvania

Larry L. Hall, Ph.D.
Toxicologist
Experimental Dosimetry Branch
Health Effects Research Laboratory
U.S. Environmental Protection Agency
Research Triangle Park, North Carolina

George S. Hawkins, Jr.
Research Chemist
Division of Cutaneous Hazards
Letterman Army Institute of Research
Presidio of San Francisco, California

John Kao, Ph.D.
Senior Investigator
Department of Drug Metabolism
Smith Kline and French Laboratories
King of Prussia, Pennsylvania

Barbara W. Kemppainen, Ph.D.
Assistant Professor
Department of Toxicology
College of Veterinary Medicine
Auburn University
Auburn, Alabama

George J. Klain, Ph.D.
Supervisory Research Chemist
Division of Cutaneous Hazards
Letterman Army Institute of Research
Presidio of San Francisco, California

Saran Kumar, Ph.D.
Associate Research Investigator
Pharmaceutical Research and
   Development
Hoffman-LaRoche, Inc.
Nutley, New Jersey

Marie Lodén, M.Sci.Pharm.
Senior Clinical Research Manager
Medical Department
ACO Läkemedel AB
Solna, Sweden

A. Waseem Malick, Ph.D.
Director
Pharmaceutical Research and
   Development
Hoffman-LaRoche, Inc.
Nutley, New Jersey

Nancy Monteiro-Riviere, Ph.D.
Assistant Professor
Department of Anatomy, Physiological
   Sciences, and Radiology
College of Veterinary Sciences
North Carolina State University
Raleigh, North Carolina

Sunil B. Patel, M.S.
Associate Scientist
Pharmaceutical Research and
  Development
Hoffman-LaRoche, Inc.
Nutley, New Jersey

David Piemontese, B.S.
Senior Lab Technician
Pharmaceutical Research and
  Development
Hoffman-LaRoche, Inc.
Nutley, New Jersey

William R. Ravis, Ph.D.
Professor and Chairman
Division of Pharmaceutics
Department of Pharmacal Sciences
School of Pharmacy
Auburn University
Auburn, Alabama

William G. Reifenrath, Ph.D.
Research Chemist
Division of Cutaneous Hazards
Letterman Army Institute of Research
Presidio of San Francisco, California

P. V. Shah, Ph.D.
Industrial Toxicologist
Hoffman-LaRoche, Inc.
Nutley, New Jersey

Kelly L. Smith
Director
Controlled Release Division
Bend Research, Inc.
Bend, Oregon

Hubert L. Snodgrass
Biologist
Toxicology Division
U.S. Army Environmental Hygiene
Agency
Aberdeen Proving Ground, Maryland

TABLE OF CONTENTS

Chapter 1

CHOICE OF MEMBRANES FOR *IN VITRO* SKIN UPTAKE STUDIES AND GENERAL EXPERIMENTAL TECHNICS

Charan R. Behl, Saran Kumar, A. Waseem Malick, Sunil B. Patel, Hing Char, and David Piemontese

TABLE OF CONTENTS

I. INTRODUCTION

Recently there has been an increased interest in the drug administration via the skin, for both local therapeutic effect on diseased skin (topical delivery) as well as for systemic delivery (transdermal delivery) of drugs. In either case, drug molecules need to be taken up by the skin. In case of topical delivery, an optimal drug residence in the skin layers (skin retention) is important whereas in the case of transdermal delivery, an optimal net transport across the skin (skin permeation) becomes an important criterion to determine the drug efficacy. Several different definitions have been reported in the literature to describe these two types of drug delivery. We would like to propose using a common term, *skin uptake*, to describe the fate of drugs upon their entry into the skin. This term has a broad definition and encompasses retention, permeation, metabolism, degradation, and binding to skin components, various activities which a drug can undergo upon entering the skin.

Regardless of the type of drug delivery, an understanding of the skin uptake properties becomes a critical factor in the development of an optimal product. In this chapter, focus will be on the skin permeability as it relates to transdermal drug delivery systems. General *in vitro* experimental considerations will be discussed with emphasis on the choice of membranes for the permeation studies.

In the development of a transdermal drug delivery system (TD-DDS), there are usually three major stages of activities involved:

- Feasibility assessment
- Formulation development
- Fabrication of device

Before launching a full scale developmental activity, it is desirable to first assess if a TD-DDS for the drug of choice is feasible. The most critical criterion in doing so is to ascertain whether or not an adequate drug delivery across the skin (skin permeation) can be achieved. This can be done by the following **four-prong approach:**

- Human *in vivo*
- Human *in vitro*
- Animal *in vivo*
- Animal *in vitro*

Of course, the most desirable approach is to do human *in vivo* experiments. Because of the difficulties such as cost, time factor, limited knowledge of the drug properties, ethical considerations, etc., it is almost impractical to begin with such an approach. Even human *in vitro* studies are difficult to conduct due to the scarcity of human skin and controlling the gender, race, site, age and skin condition of the donor. Therefore, a lot of reliance is put on the use of animal models. Since the *in vivo* experiments are more time consuming, cumbersome and expensive, the bulk of skin permeation studies are carried out *in vitro* with animal skins. This trend is strengthened and encouraged by the fact that for almost all drugs, the stratum corneum is the rate controlling membrane.[1-10] Since the stratum corneum is made of keratinized dead cells, it is widely believed and accepted that the *in vitro* skin permeation results are a good representation of *in vivo* situation.[11-35]

II. CHOICE OF MEMBRANES

A. TYPES OF MEMBRANES AVAILABLE
1. Human Skin

The effects of anatomical site, gender, age, and race on the skin permeability have been

TABLE 1
List of Various Animals Which
Have Been Used in Skin
Permeation Studies

Animal	Ref.
Hairless mouse	8, 9, 44, 45, 64
Swiss (furry) mouse	64, 65, 66
Athymic nude mouse	33
Furry rat	67, 68, 69
Fuzzy rat	70, 71
Nude rat	72, 73, 74
Hairless rat	72, 75, 76
Hairless guinea pig	77, 78
Guinea pig	6, 21, 72, 79
Weanling yorkshire pig	6, 35
Yucatan miniature pig	80
Mini pig	35, 66, 68
Micro pig	81
Domestic pig	82
Cat	6, 83
Dog	6, 83, 84
Mexican hairless dog	85
Hairless dog	33
Syrian golden hamster	86
Sheep	87
Chimpanzee	6
Rhesus monkey	41, 42, 54, 88, 89
Squirrel monkey	66
Rabbit	72, 88
Goat	6, 83
Horse	6
Snake (shed skin)	90

reported in numerous publications both for human and animal skins.[1,21,36-60] Additionally, in some cases, even the race has been shown to be an important factor.[61-63] Unfortuanately, one does not have a choice in selecting human skin of desired characteristics because of practical difficulties in obtaining skin specimens. The most readily available and frequently used skin sites are abdominal and thigh from autopsies of older humans. With the given limitations, care should still be taken to control these variables. Because of the rate controlling properties of the stratum corneum, most researchers tend to use only the epidermis. But, depending upon the need of the study, whole skin, epidermis, stripped skin, or dermis can be used. In the case of whole skin, the fat should be removed carefully and completely. If the skin is sectioned, the method chosen should be such that it does not cause an artefact in the data. Some sort of methods validation should be done.

2. Animal Skins

Over the past many years, a large number of different animal skins have been tested as possible models for human skin. These animals are listed in Table 1.[6,8,9,21,33,35,41,42,44,45,54,64-90] An attempt has been made to include some selected references for each animal type. The list is by no means a complete compilation of all different kinds of animal skins utilized. It, however, gives a clear indication of the diversity in researchers' choice of *in vitro* models presumably mimicking the human skin permeability. These animal skins can be divided into three major types:

TABLE 2
Summary of Artificial Membranes Used in
Permeation Studies

Membrane type	Ref.
Silastic® (polydimethylsiloxane)	114—130
Cellulose acetate	131—138
Polyurethane	139, 140
Supor® (modified polysulfone)	141
Zeolite (aluminosilicates)	142
Multimembrane system (cellulose acetate: Silastic® cellulose acetate)	143
Diaflo-ultrafiltration membrane	144
Miscellaneous: collagen, egg shell	145, 146
Liquid membranes (organic liquids, e.g., isopropyl myristate soaked on filter membranes.)	147—152

- Furry skins
- Fuzzy skins
- Hairless skins

The reader is advised to get a thorough understanding of the anatomical and histological characteristics of these skins by reading appropriate references.[91-113]This knowledge is helpful in interpreting the skin permeation data. A review of the skin uptake literature indicates that different investigators have found that a variety of animal skins resemble the human skin permeability. Different permeants, permeation apparatus and methodologies were used in these studies. Also, the human skin used for comparison purposes came from different anatomical sites and was obtained from varying sources. It appears that the criterion of establishing a correlation between the animal and human skin was the permeation rates. This criterion is grossly in error especially when there are substantial differences within the human skin. For example the skins from different sites of the human body have widely different permeabilities. Also, skin permeability varies as a function of individuals, race, gender, age, overall skin conditioning, and state of skin, i.e., diseased or normal. It is, therefore naive to expect the skin permeability of an animal such as the mouse (skin thickness = 200 μm) to be comparable to that of human skin (skin thickness = 1 to 3mm). Human skin, for most part, is not a furry skin. Therefore, why should the skin of furry animals which have different skin texture and degree of keratinization, have permeability comparable to that of human skin? The question can be further extended: why should the skin permeability of a given animal (e.g., mouse, rat, guinea pig, pig, cat, dog, rabbit, etc.) be comparable to the *in vivo* human skin permeability? Again, to which anatomical site, race, gender, age, etc., of the human skin should one compare the animal data. For a comparison of animal skin permeability to that of human skin, the mechanistic aspects of the permeation process should be used as the criterion.

3. Artificial Membranes

Some researchers have explored the use of synthetic membranes in the skin permeation studies (Table 2). Three possible reasons for using these membranes are as follows.

a. To Understand the Mechanistic Aspects of Skin Uptake Processes

At least three types of synthetic membranes have been used for a better understanding of the skin uptake processes; Silastic® (polydimethyl siloxane),[114-130] cellulose acetate,[131-138] and polyurethane,[139,140] membranes. Some of the other less extensively studied membranes include Supor® (modified polysulfone),[141] Zeolites,[142] multimembrane systems,[143] and Diaflo-ultrafiltration membrane.[144] Some membranes of biological origin, e.g., collagen and egg shell, have also been investigated.[145,146] There are some studies where liquid membranes were used in different ways.[147-152] In some cases, special three compartment apparatus were used to study the rate of permeation across organic liquid layers whereas in other studies filter paper soaked in organic liquids was used as the membrane.

In choosing a membrane for mechanistic understanding of skin permeation processes, the membrane selected should be such that the permeation pathways (one or more) between the two membranes are common. Only two artificial membranes have been evaluated in a systematic manner with this criterion in mind, Silastic[130] and polyurethane.[140] Actually it is a difficult area of research. Skin is a complex organ which is multilayered and heterogenous and presents a very sophisticated barrier system to drug transport. The synthetic membranes used to date are simple homogenous monolayer systems. Any comparison of the permeability data with the skin must therefore be done with extreme caution. One important utility of artificial membranes is to differentiate the physicochemical (thermodynamic) dependent properties of permeation from the permeation processes which depend on the biological properties of the skin. For example the pure thermodynamic effects of drug solubility, partition coefficient, pH, drug-excipient interaction, etc., can be understood by the use of artificial membranes. However, the factors which affect the skin and thus alter its permeability or drug diffusivity through it, cannot be judged by using synthetic membranes. For example, most enhancers accelerate drug permeation primarily by causing some changes in the skin and therefore their performance cannot be adequately studied and understood from the artificial membranes' behavior. Unfortunately, almost all excipients used in topical and transdermal products are taken up by the skin and permeate through it. Depending on the nature of the excipients used, minor to severe skin effects can be noted. This makes the use of artificial membranes less valuable because these skin effects cannot be predicted from such membranes. For example certain organic solvents such as ethanol alter skin permeability by affecting the lipids. Artificial membranes are devoid of lipids and thus are not useful to predict such an effect.

b. As an Alternate to Skin

Due to the difficulties and problems involved in using skin, both *in vitro* and *in vivo*, there is a genuine desire to find an alternate membrane to be used routinely in screening drugs and formulations for skin permeation profiles. Research in this area is ongoing in various laboratories. It is our opinion that it is not feasible to find such an alternate membrane to the skin in the forseeable future. Due to the complexities of skin, the absolute skin permeabilities or even rank orders cannot be predicted adequately from the synthetic membrane data. Efforts to prepare artificial skin by growing epidermal cells on synthetic collagenous sheets are on-going in at least one company, Marion laboratories, Kansas City, Kansas. This artificial skin is being developed for grafting on wounds. When available, it will be interesting to compare its permeation profiles to those of the human skin. Even the artificial skin, in our opinion, will not be able to predict a complete profile of the human skin permeability.

c. Quality Control Tool to Ascertain Batch to Batch Uniformity

In a typical TD-DDS, the rate controlling mechanism lies in the drug release from the device and does not depend on the skin permeation. Therefore, it should be possible to use an alternate membrane to the skin to determine batch to batch uniformity. In fact there should be no need to use any membrane to study the release from these devices as the drug release can be done directly in a suitable medium. This seems to be the method of choice at this time both in the research laboratories as well as at the Food and Drug Administration laboratories.[153]

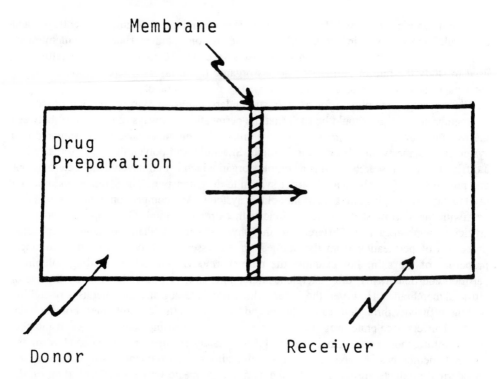

FIGURE 1. A schematic representation of a two compartment horizontal permeation cell. The contents of each compartment are stirred and the whole unit is maintained at a constant temperature.

B. PERMEABILITY OF TEST COMPOUNDS THROUGH DIFFERENT MEMBRANES

There is an abundance of *in vitro* permeation data available in the literature where different kinds of membranes were used. In these studies, different types of compounds were used and the membrane permeability was determined by using different kinds of apparatus and formulations. Therefore, it is difficult to pool these data and draw a common conclusion about the suitability of other membranes to human skin. However, some researchers have focused on a common series of permeants, *n*-alkanols and a common type of two compartment horizontal permeation cell. *n*-Alkanols are simple, small, stable, and neutral molecules and are suitable model compounds to elucidate permeation mechanisms. A substantial amount of data has been gathered using human skin, animal skins, and synthetic membranes. These data will be presented and discussed in this section.

The permeation data presented here were obtained from the pseudo steady-state slopes of drug transport across a membrane sandwiched between two cell-halves of a permeation apparatus (Figure 1). The permeability coefficients were calculted from the following simplified Fick's diffusion equation:

$$P = J/(\Delta C \cdot A) \tag{1}$$

$$P = (dc/dt)/(C_d \cdot A) \tag{2}$$

where P = permeability coefficient (cm/h); J = flux (mg/h); ΔC = drug concentration gradient between the two cell halves (mg/cm^3); dc/dt = slope of the steady-state portion of amount vs. time plot (mg/h); A = skin area exposed to drug solution (cm^2); and C_d = drug concentration in the donor compartment and equals to ΔC because of low concentration in the receiver compartment.

The permeability coefficient is constant for a given permeant under a given set of experimental conditions and depends on

- Diffusion coefficient of drug (D)
- Partition coefficient of drug (K)
- Membrane thickness (h)

in accordance to:

$$P = DK/h \qquad (3)$$

The diffusion coefficient is also a constant property of a drug and depends on the following factors:

- State of permeant, e.g., gas, liquid, or solid
- Temperature
- Viscosity
- Molecular size

The partition coefficient is the ratio of relative equilibrium drug concentration in the membrane and the formulation. In most cases, the partition coefficient is the primary factor in determining drug permeation across the skin. The membrane thickness is the thickness of the rate controlling layer of the skin and not the whole skin. It is important to understand this aspect so that an appropriate mechanistic interpretation of the permeation data can be made. In most cases, the rate controlling layer is the stratum corneum which is a thin, heterogenous, and multicomponent membrane. The precise composition, texture, appendageal distribution, water/lipid/protein contents and compaction of the stratum corneum varies amongst animal species, animal to human, and amongst humans from site to site and with race, age, and gender. This can make the permeability to membrane thickness relationship a variable factor. In other words, a thinner stratum corneum does not necessarily mean that it is more permeable. Therefore, one needs to be careful in interpreting or extrapolating the permeation data based only on the membrane thickness.

In the following section, the n-alkanol permeation data for different membranes will be presented and discussed. An attempt is made to present all available data. In some cases, a complete set of alkanol permeation rates is not available. Research with these test permeants is on-going in our laboratories and elsewhere.

1. Human Skins

Table 3 contains a summary of human skin permeability coefficients of n-alkanols.[1] Scheuplein and co-workers studied these model compounds to understand the permeation pathways in the skin. When these data are plotted semilogarithmically as a function of the alkyl chain length (Figure 2), a sigmoidal shaped curve was obtained. The steep rising portion of the curve was attributed to the lipoidal partitioning process. From this portion, the slope was computed and was termed the π-value (Table 3). A π-value of 0.31 was obtained for the human skin.

2. Animal Skins

The n-alkanol permeation data for hairless mouse,[44] swiss (furry) mouse,[154] fuzzy rat,[70] nude rat,[73] hairless rat,[75] furry rat,[69] and hairless guinea pig[78] skins are summarized in Table 3. In most cases data are reported for water and alkanols ranging from methanol to octanol. Data for hairless mouse skin are provided at three different ages, 5, 25, and 210d. The

TABLE 3
Summary of *In Vitro* n-Alkanol Permeability Coefficients (*p* Values) for Human and Animal Skins

		$p \times 10^3$ (cm/h) Skin membrane								
		Animal skin								
		Hairless mouse; age (d)								Hairless
Permeant	Human skin	5	25	210	Furry mouse	Hairless rat	Nude rat	Fuzzy rat	Furry rat	guinea pig
Water	0.50	0.80	1.90	0.70	10.1	—	2.40	—	—	0.94
Methanol	0.50	1.20	3.40	1.00	9.50	—	4.00	0.64	5.60	1.21
Ethanol	0.80	0.90	3.50	0.90	4.70	1.00	4.10	0.72		1.82
Propanol	1.40	—	—	—	—	—	—	—	—	—
Butanol	2.50	4.20	12.7	4.10	14.9	2.40	11.0	2.79	5.80	4.61
Pentanol	6.00	—	—	—	14.3	—	—	—	—	—
Hexanol	13	20.5	54.1	17.8	19.0	—	41.0	8.71	9.60	20.0
Heptanol	32	—	—	—	25.5	—	—	—	—	—
Octanol	52.0	52.5	118.4	14.0	25.9	30.0	61.6	15.7	9.48	46.1
Nonanol	60.0	—	—	—	—	—	—	—	—	—
π-value[a]	0.31	0.34	0.30	0.32	0.15	0.25[b]	0.25	0.27	0.11[c]	0.26
Ref.	1	44	44	44	154	75	73	70	69	78

[a] In most cases, π-value was computed as the slope of the steep rise portion of log P vs. alkyl chain length plots using data for ethanol through hexanol.

[b] Due to limited data available, π-value was computed from data for ethanol, butanol and octanol.

[c] Due to limited data available, π-value was computed from data for only butanol and hexanol.

respective π-Values are also provided for these skins. Figure 3 contains semilogarithmic plots of *p* values vs. the alkyl chain length. One subplot is provided for each of the skin types. The general shape of these plots is mutally comparable. However, there are some subtle differences in the profiles and also the π-values differ for some of these skins. The furry skins appear to show much lower partitioning dependent permeation than the fuzzy and hairless skins. This is seen in the permeation profiles as well as in lower π-values.

3. Artificial Membranes

The alkanol permeation data for two synthetic membranes, Siliastic® and polyurethane are provided in Table 4.[130,140] These data are presented for three thicknesses of Silastic®. The permeation profiles of these data are presented in Figure 4. Note that the Silastic® data for different thicknesses converge at octanol indicating an aqueous boundary layer controlled permeation process for octanol (and higher alkanols). The π-values were computed from these plots and are reported in Table 4.

The permeability of alkanols through synthetic membranes is sensitive to the donor compartment medium. This sensitivity is noted from data presented in Table 4 on the permeability coefficients of *n*-alkanols through Silastic® and polyurethane membranes as a function of donor medium. For this purpose data from water (buffer) and Azone/Tween 80/ water emulsion are presented and compared. As can be seen from graphical presentation of these data (Figure 4), by changing the donor medium, there is a major effect on the permeation profiles. These profiles change from positive slope to a negative slope. This change is most likely due to changed partitioning behavior of the permeants and it emphasizes the effects of formulation variables on membrane permeability.

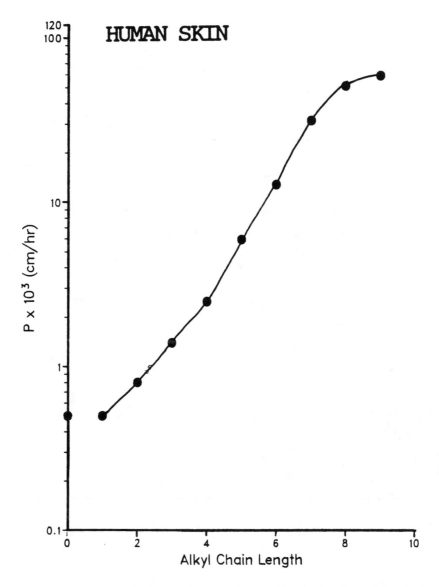

FIGURE 2. Semilogarithmic plot of *in vitro* permeability coefficients (P) vs. the alkyl chain length of n-alkanols for human skin.[1] Data at zero alkyl chain length is for ³H-water.

C. CRITERIA IN CHOOSING MEMBRANES
1. Skin Permeation Profiles

The choice of a certain membrane for *in vitro* permeation studies should be such that the data are mechanistically comparable to the human skin permeability profiles. Note that it is not necessary that the absolute skin permeability of a given membrane be the same as that of human skin. An evaluation of *n*-alkanol permeation data through human skin[1] and animal skin[44,69,70,73,75,78,154] as well as other biological membranes such as the gastrointestinal tract,[155] rabbit vaginal epithelium,[156] rhesus monkey vaginal epithelium[157] indicates a general and common pattern of log (P) vs. the alkyl chain length profile. As shown in Figure 5,[158] there are three primary transport pathways through biological membranes which are as follows.

● *Aqueous pore type pathway* which is represented by the lower plateau (Region I) and

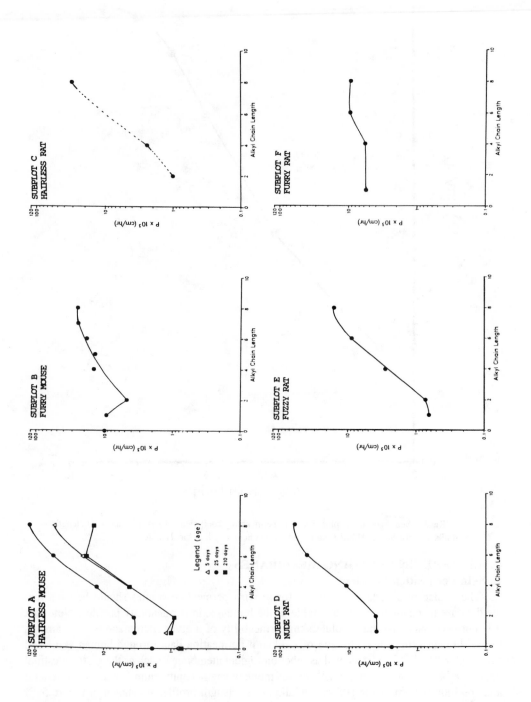

11

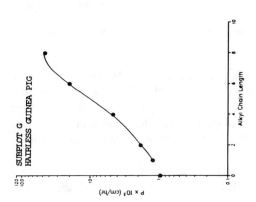

FIGURE 3. Semilogarithmic plots of *in vitro* permeability coefficients (P) vs. the alkyl chain length of n-alkanols for animal abdominal skins. Data at zero alkyl chain length are for ^{3}H-Water. Subplot A = Hairless mouse, 5, 25 and 210 d old;[44] subplot B = Swiss (furry) mouse;[154] subplot C = Hairless rat;[75] subplot D = Nude rat;[73] subplot E = Fuzzy rat;[70] subplot F = Furry rat;[69] subplot G = Hairless guinea pig;[78]

TABLE 4
Summary of *In Vitro* n-Alkanol Permeability Coefficients (*p*-Values) for Silastic® and Polyurethane Membranes from Different Donor Media

	$p \times 10^3$ (cm/h)					
	Silastic®; donor medium				Polyurethane; donor medium	
	0.9% NaCl aqueous soln;(μm)			Azone: Tween 80:	0.9% aqueous	Azone: Tween 80:
Permeant	40	80	250	water emulsion	NaCl soln	water emulsion
Methanol	—	32.8	18.3	7.09	0.54	8.64
Ethanol	94.9	59.9	34.8	—	0.55	—
Butanol	334.7	202.7	104.3	3.06	1.17	8.95
Pentanol	628.8	427.8	262.2	—	—	—
Hexanol	978.7	764.3	530.6	1.44	13.2	6.49
Heptanol	1196	1216	852.7	—	—	—
Octanol	1311	1260	1200	0.32	37.0	5.71
π-value[a]	0.26	0.28	0.29	Negative slope	0.35[b]	Negative slope
Ref.	130	130	130	140	140	140

[a]   π-value was calculated as the slope of the steep rise of log P vs. the alkyl chain length plots using data for ethanol through hexanol.
[b]   This is the best possible estimate at this time. Due to the shape of the plot, additional data for more alkanols are needed to compute a more precise π-value

for the skin it occurs through the appendages, e.g., hair follicles and sweat glands and imperfections in the surface layer of the cornified cells of the stratum corneum.

- *Lipid partitioning pathway* which is represented by the steep rising portion of the curve (Region II) and occurs as a result of permeant partitioning into the lipoidal components of the stratum corneum. The slope of this region is termed the π-value and is an important parameter.

- *Aqueous tissue pathway* which is represented by the second plateau region (Region III) and the permeation is primarily controlled by the resistance offered by the aqueous tissue of the skin.

Any membrane which provides these three permeation pathways as they occur in the human skin, should be an appropriate model membrane for *in vitro* skin permeation studies.

For Silastic® membranes, there are only two pathways available, partition dependent pathway and an aqueous boundary layer controlled pathway represented by an upper plateau (Figure 4). The polyurethane membrane also shows two pathways, an aqueous pore type pathway and an apparent partition dependent pathway. It also provides an indication of a third pathway which is not explainable at this time. The pore type pathway is consistent with the texture of this membrane. The permeability of alkanols through this 40-μm thick membrane is generally lower than that through 50, 80, or 250-μm thick Silastic® membrane. The polyurethane membrane permeability of octanol is far below the permeability level where the permeation process becomes aqueous boundary layer controlled. Therefore, the boundary layer controlled mechanism is not relevant to this membrane.

With the above discussion, we can now compare the various membranes for their suitability to human skin. It appears that in general the furry skins do not show permeation profiles and π-values which are comparable to those of the human skin. This observation is consistent with widely different anatomical and histological features of the furry skins. Therefore, furry animal skin does not provide a suitable model for *in vitro* skin permeation studies. The fuzzy, nude or hairless animal skins appear to provide the permeation profiles

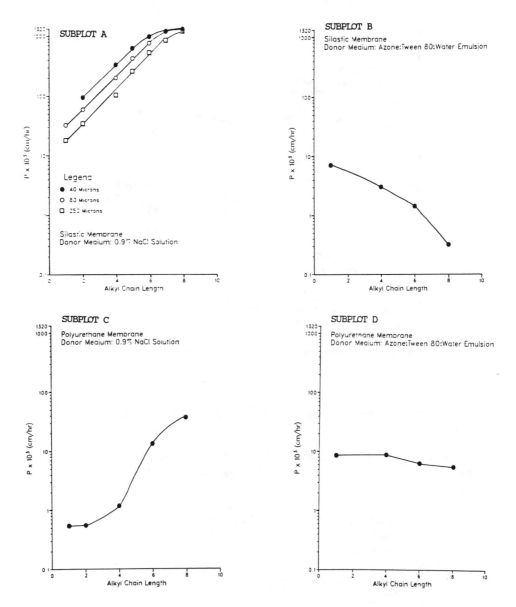

FIGURE 4. Semilogarithmic plots of permeability coefficients (P) vs. the alkyl chain length of n-alkanols for artificial membranes. Subplot A = Silastic® of 40, 80, and 250 μm thicknesses. Donor medium was aqueous solution;[130] subplot B = Silastic® of 250 μm thickness. Donor medium was Azone/Tween 80/Water emulsion;[140] subplot C = Polyurethane. Donor medium was aqueous solution;[140] subplot D = Polyurethane. Donor mediums was Azone/Tween 80/Water emulsion;[140]

and the π-values which are comparable to those of the human skin. Such skins are generally good models for *in vitro* skin permeation studies. The fuzzy rat and nude rat skins contain a lot of scattered hair which need to be clipped prior to the experiments. These skins, in our experience, are not particularly desirable membranes for *in vitro* permeation studies. The skin surface of hairless rodents is smooth, fully devoid of hair and easy to handle in the laboratory. Among these animal species, the hairless mouse skin has been most extensively used. But this trend is now changing. This skin is so thin that it substantially overpredicts the vehicle and enhancer effects.[159] Also, when iontophoretic experiments are carried out, the hairless mouse skin suffers extensive damage even at low voltage/current applica-

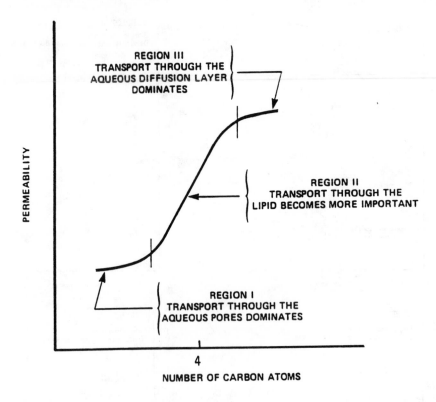

FIGURE 5.   A schematic representation of three pathways of permeation through biological membranes.[157,158]

tions.[160] The use of hairless guinea pig skin is becoming popular amongst various researchers including ourselves.[77,78,113] This animal model has the following advantages over other rodents:

- The skin is smooth and fully devoid of hair
- Large enough animal to carry out *in vitro, in situ,* or *in vivo* experiments
- Easy to handle and to excise skin
- Provides sufficient skin for several *in vitro* experiments (up to 16) per animal
- Skin permeation pathways are comparable to those of the human skin

Care should be taken so that the hairless guinea pigs do not gain excess weight. If over weight, the skin accumulates extra fat on the underside of the skin which is difficult to remove. Also, if the animals are too old, e.g., over 6 months age, the skin tends to build up excess fat. In our experience animals of 2 to 6 months age are most suitable. Hairless guinea pig at this time appears to be a very promising laboratory animal model for *in vitro* and *in vivo* skin permeation studies.

2. Metabolic (Degradation) Considerations

Most drugs are sensitive to enzymatic metabolism, degradation, and isomerization in the skin. Whenever these factors are important for the drugs in question, one should select an animal model more cautiously. Also, the suitablity of *in vitro* procedures vs. *in vivo* experiments should be studied. Sometimes, it may be necessary to carry out *in vivo* studies to get more appropriate metabolic data.

3. Availability of Membranes

The membrane chosen for skin permeation studies should be readily available from

TABLE 5
Sources of Recommended Membranes for Skin Uptake Studies

Membrane	Source
Human skin (abdominal or thigh)	1. Pathology department at local hospital (fresh skin from autopsy cases) 2. Local medical examiners' office (fresh skin) 3. Local branch of organ bank (stored and frozen skin) 4. From surgical procedures at local hospitals (fresh skin)
Hairless guinea pig skin	Charles River Laboratories, 251 Ballardvale Street, Wilmington, MA, 01887-0630
Silastic® (polydimethylsiloxane) 0.01 in. thick membrane	Dow Corning Corp. Box 1767 Midland, MI, 48640
Polyurethane membrane	Norwood Industries, Inc. A subsidiary of Seton Co. 100 North Morehall Rd., Malvern, PA, 19355

reliable sources (Table 5). In as much as possible, the type and source of the membrane should be kept constant. Often changing the quality or supplier of membranes has effects on the experimental data. This is true not only for synthetic membranes but also for animal skins. For example the hairless mouse skin obtained from one supplier may have different anatomical and histological properties than the animal obtained from another supplier. This may be so due to differences in the breeding, feeding and maintenance procedures. Also the starting genetic material of animals may be quite different between suppliers. It is desirable that the animal be purchased from the nearest possible location of the supplier. It avoids longer transportation of animals which can affect the quality of the skin.

4. Cost of Membranes

Cost is also an important factor in choosing membranes. If an expensive membrane is selected, it often restricts the volume of data which can be collected thus prohibiting carrying out a thorough study. On the other hand, the quality should not be compromised to a point where it reflects on the value of experimental observations.

D. COMMERCIAL SOURCES OF MEMBRANES

A list of some suppliers of skins (human and animal) and two synthetic membranes is provided in Table 5. Emphasis should be placed on the reliability of the source with regard to assurance of continuing supply and maintenance of quality. It is useful to inform the supplier of the nature of your research and stress the importance of membrane integrity especially the stratum corneum.

REFERENCES

1. **Scheuplein, R. J. and Blank, I. H.,** Permeability of the skin, *Physiol. Rev.,* 51, 702, 1971.
2. **Scheuplein, R. J.,** Mechanisms of percutaneous absorption. I. Routes of penetration and influence of solubility, *J. Invest. Dermatol.,* 45, 334, 1965.
3. **Scheuplein, R. J.,** Mechanisms of percutaneous absorption. II. Transient diffusion and relative importance of various routes of penetration, *J. Invest. Dermatol.,* 48, 79, 1967.
4. **Stoughton, R. B.,** Animal models for *in vitro* percutaneous absorption, in *Animal Models in Dermatology,* Maibach, H. I., Ed., Churchill-Livingstone, New York, 1975, 121.
5. **Kligman, A. M.,** The biology of the stratum corneum, in *The Epidermis,* Montagna, W. and Lobitz., W. C., Eds., Academic Press, New York, 1964, chap. 20.

6. **Marzulli, F. N., Brown, D. W. C., and Maibach, H. I.,** Techniques for studying skin penetration, *Toxicol. Appl. Pharmacol. Suppl.,* 3, 76, 1969.

7. **Wurster, D. E. and Kramer, S. F.,** Investigation of some factors influencing percutaneous absorption, *J. Pharm Sci.,* 50, 288, 1961.

8. **Flynn, G. L., Durrheim, H. H., and Higuchi, W. I.,** Permeation of hairless mouse skin. II. Membrane sectioning techniques and influence on alkanol permeabilities, *J. Pharm. Sci.,* 70, 52, 1981.

9. **Behl, C. R., Barrett, M., Flynn, G. L., Kurihara, T., Walters, K. A., Gatmaitan, O. G., Harper, N., Higuchi, W. I., Ho, N. F. H., and Pierson, C. L.,** Hydration and percutaneous absorption. III. Influences of stripping and scalding on hydration alterations of the permeability of hairless mouse skin to water and n-alkanols, *J. Pharm. Sci.,* 71, 229, 1982.

10. **Idson, B.,** Percutaneous absorption, *J. Pharm Sci.,* 64, 901, 1975.

11. **Burch, G. E. and Winsor, T.,** Rate of insensible perspiration locally through living and dead human skin, *Arch. Intern. Med.,* 74, 437, 1944.

12. **Feldman, R. J. and Maibach, H. I.,** Absorption of some organic compounds through the skin in man, *J. Invest. Dermatol.,* 54, 399, 1970.

13. **Franz, T. J.,** Percutaneous absorption. On the relevance of *in vitro* data, *J. Invest. Dermatol.,* 64, 190, 1975.

14. **Franz, T. J.,** The finite dose technique as a valid *in vitro* model for the study of percutaneous absorption, *Curr. Probl. Dermatol.,* 7, 58, 1978.

15. **Sekura, D. and Scala, J.,** The percutaneous absorption of alkyl methyl sulfoxides, in *Pharmacology and the Skin,* Montagna, W., Van Scot, E., and Stoughton, R., Eds., Appleton-Century-Crafts, New York, 1972, 257.

16. **Creasey, N. H., Battensby, J., and Fletcher, J. A.,** Factors affecting the permeability of skin, *Curr. Probl. Dermatol.,* 7, 95, 1978.

17. **Bronaugh, R. L., Stewart, R. F., Congdon, E. R., and Giles, A. L., Jr.,** Methods for *in vitro* percutaneous absorption studies. I. Comparison with *in vivo* results, *Toxicol. Appl. Pharmacol.,* 62, 474, 1982.

18. **Bronaugh, R. L., Stewart, R. F., Congdon, E. R., and Giles, A. L., Jr.** Methods for *in vitro* percutaneous absorption studies. II. Animal models for human skin, *Toxicol. Appl. Pharmacol.,* 62, 481, 1982.

19. **Astley, J. P. and Levine, M.,** Effect of dimethyl sulfoxide on permeability of human skin *in vitro,* J. Pharm. Sci., 65, 211, 1976.

20. **Ostrenga, J. C., Steinmetz, C., and Poulsen, B.,** Significance of vehicle composition. I. Relationship between topical vehicle composition, skin penetrability and clinical efficacy, *J. Pharm. Sci.,* 60, 1175, 1971.

21. **Tregear, R. T.,** Factors affecting the skin permeability. Molecular movement, in *Theoretical and Experimental Biology,* Vol. 5, Academic Press, London, 1966, 12.

22. **Blank, I. H.,** Factors which influence the water content of the stratum corneum, *J. Invest. Dermatol.,* 18, 433, 1952.

23. **Baker, H. and Kligman, A. M.,** Measurement of transepidermal water loss by electrical hygrometry. Instrumentation and responses to physical and chemical insults, *Arch. Dermatol.,* 96, 441, 1967.

24. **Bettley, F. R. and Grice, K. A.,** A method for measuring the transepidermal water loss and a means of inactivating sweat glands, *Br. J. Dermatol.,* 77, 627, 1965.

25. **Mali, J. W.,** The transport of water through the human epidermis, *J. Invest. Dermatol.,* 27, 451, 1956.

26. **Onken, H. D. and Moyer, C. A.,** The water barrier in human epidermis. Physical chemical nature, *Arch. Dermatol.,* 87, 584, 1963.

27. **Rosenberg, E. W., Blank, I. H., and Resnik, S.,** Sweating and water loss through the skin. Studies using an electrical humidity sensor, *JAMA,* 179, 809, 1962.

28. **Rutter, N. and Hull, D.,** Response of term babies to a warm environment, *Arch. Dis. Child.,* 54, 178, 1979.

29. **Orsmark, K., Wilson, D., and Maibach, H. I.,** *In vivo* transepidermal water loss and epidermal occlusive hydration in newborn infants: anatomical region variation, *Acta Derm. Venerol. (Stockh),* 60, 403, 1980.

30. **Oishi, H., Ushio, Y., Narahara, K., and Takehara, M.,** Effect of vehicles on percutaneous absorption. I. Characterization of oily vehicles by percutaneous absorption and trans-epidermal water loss test, *Chem. Pharm. Bull. Tokyo,* 24, 1765, 1976.

31. **Lamke, L. O., Nilsson, G. E., and Reithner, H. L.,** Insensible perspiration from the skin under standardized environmental conditions, *Scand. J. Clin. Lab. Invest.,* 37, 325, 1977.

32. **Elias, P. M., Brown, B. E., and Ziboh, V. A.,** The permeability barrier in essential fatty acid deficiency: evidence for a direct role of linoleic acid in barrier function, *J. Invest. Dermatol.,* 74, 230, 1980.

33. **Reifenrath, W. G., Chellquist, E. M., Shipwash, E. A., Jederberg, W. W., and Kruger, G. G.,** Percutaneous absorption in hairless dog, weanling pig and grafted athymic nude mouse: evaluation of models for predicting skin penetration in man, *Br. J. Dermatol.,* 3, (Suppl. 27), 123, 1984.

34. **Reifenrath, W. G., Mershon, M. M., Brinkley, F. B., Miura, G. A., Broomfield, C. A., and Cranford, H. B.,** Evaluation of diethyl malonate as a simulant for 1,2,2-trimethylpropyl methylphosphonofluoridate (Soman) in shower decontamination of the skin, *J. Pharm. Sci.,* 73, 1388, 1984.

35. **Reifenrath, W. G. and Hawkins, G. S.,** The weanling yorkshire pig as an animal model for measuring percutaneous penetration, in *Swine in Biomedical Research,* Vol 1, Tumbleson., M. E., Ed., Plenum, New York, 1986, 673.

36. **Nachman, R. L. and Ester, N. B.,** Increased skin permeability in preterm infants, *J. Pediatr.* 79, 628, 1971.

37. **Feiwel, M.,** Percutaneous absorption of topical steroids in children, *Br. J. Dermatol.* 81, Suppl. 4, 113, 1969—1970.

38. **Cronin, E. and Stoughton, R. B.,** Percutaneous absorption, regional variations and effects of hydration and epidermal stripping, *Br. J. Dermatol.,* 74, 265, 1962.

39. **Smith, J. G., Fischer, R. W., and Blank, I. H.,** The epidermal barrier — a comparison between scrotal and abdominal skin, *J. Invest. Dermatol.,* 36, 337, 1961.

40. **Marzulli, F. N.,** Barriers to skin penetration, *J. Invest. Dermatol.,* 39, 387, 1962.

41. **Feldman, R. J. and Maibach, H. I.,** Regional variations in percutaneous absorption of ^{14}C-cortisol in man, *J. Invest. Dermatol.,* 48, 181, 1967.

42. **Maibach, H. I., Feldman, R. J., Milby, T. H., and Serat, W. F.,** Regional variations in percutaneous penetration in man — pesticides, *Arch. Environ. Health,* 23, 208, 1971.

43. **Shuster, S., Black, M. M., and Mevitie, E.,** The influence of age and sex on skin thickness, skin collagen and density, *Br. J. Dermatol.,* 93, 639, 1975.

44. **Behl, C. R., Flynn, G. L., Kurihara, T., Smith, W. M., Bellantone, N. H., Gatmaitan, O., Higuchi, W. I., and Ho, N. F. H.,** Age and anatomical influences on alkanol permeation of skin of male hairless mouse, *J. Soc. Cosmet. Chem.,* 35, 237, 1984.

45. **Cline, S., Behl, C. R., and Flynn, G. L.,** Skin permeability studies in hairless and swiss mice. Effects of age and alkyl chain length on *in situ* permeabilities of n-alkanols, Presented at the 39th National Meeting of the Academy of Pharmaceutical Sciences, A.Ph.A., Minneapolis, MN, October 20 — 24, 1985, Basic Pharmaceutics Abstract # 27.

46. **Behl, C. R., Flynn, G. L., Kurihara, T., Smith, W. M., Bellantone, N. H., Gatmaitan, O., Pierson, C. L., Higuchi, W. I., and Ho, N. F. H.,** Aging and anatomical site influences on the permeation of water and n-alkanols through female hairless mouse skin, Presented at the 131st American Pharmaceutical Association Annual Meeting, Montreal, May 5 — 10, 1984. Basic Pharmaceutics Abstract # 2.

47. **Harpin, V. and Rutter, N.,** Percutaneous alcohol absorption and skin necrosis in a preterm infant, *Arch. Dis. Child.* 57, 477, 1982.

48. **Bronaugh, R. L. and Maibach, H. I.,** *In vitro* percutaneous absorption, in *Dermatotoxicology,* Marzulli, F. N and Maibach, H. I., Eds., Hemisphere, Washington, D.C., 1983, 117.

49. **Hammarlund, K., Nilsson, G. E., Oberg, P. A., and Sedin, G.,** Transepidermal water loss in newborn infants. I. Relation to ambient humidity and site of measurement and estimation of transepidermal water loss, *Acta. Paediatr. Scand.,* 66, 552, 1977.

50. **Bronaugh, R. L., Stewart, R. F., and Congdon, E. R.,** Differences in permeability of rat skin related to sex and body site, *J. Soc. Cosmet. Chem.,* 34, 127, 1983.

51. **Idson, B.,** Biophysical factors in skin penetration, *J. Soc. Cosmet. Chem.,* 22, 615, 1971.

52. **Tregear, R. T.,** Relative permeability of hair follicles and epidermis, *J. Physiol.,* 156, 307, 1961.

53. **Scheuplein, R. J.,** Site variations and permeability, in *The Physiology and Pathophysiology of the Skin,* Vol. 5, Jarrett., A., Ed., Academic Press, New York, 1978, 1731.

54. **Wester, R. C., Noonan, P. K., and Maibach, H. I.,** Variation in percutaneous absorption of testosterone in the rhesus monkey due to anatomical site of application and frequency of application, *Arch. Dermatol. Res.,* 267, 229, 1980.

55. **Shaw, J. E. and Chandrasekharan, S. K.,** Transdermal therapeutic systems, in *Drug Absorption,* Prescott, L. F. and Nimov, W. S., Eds., Academic Press, New York 1981, chap. 18.

56. **Rougier, A. R., Lotte, C., Corcuff, P., and Maibach, H. I.,** Relationship between skin permeability and corneocyte size according to anatomic site, age, and sex in man, *J. Soc. Cosmet. Chem.,* 39, 15, 1988.

57. **Christopher, E. and Kligman, A. M.,** Percutaneous absorption in aged skin, in *Advances in the Biology of the Skin,* Vol. 6, Montagna, W., Ed., Pergamon Press, New York, 1965, 163.

58. **Wildnauer, R. H. and Kennedy, R.,** Transepidermal water loss of human newborns, *J. Invest. Dermatol.,* 54, 483, 1970.

59. **Wester, R. C., Noonan, P. K., Cole, M. P., and Maibach, H. I.,** Percutaneous absorption of testosterone in the newborn rhesus monkey: comparison to the adult, *Pediatr. Res.,* 11, 737, 1977.

60. **Shuman, R. M., Leech, R. W., and Alvord, E. C.,** Neurotoxicity of hexachlorophene in humans. II. A clinicopathological study of 46 premature infants, *Arch. Neurol.,* 32, 320, 1975.

61. **Weingand, D. A., Haygood, C., Gaylor, J. R., and Anglin, J. H., Jr.,** Racial variations in the cutaneous barrier in current concepts, in *Cutaneous Toxicity,* Drill, V. A. and Lazar, P., Eds., Academic Press, New York, 1980, 221.

62. **Stoughton, R. B.,** Bioassay of antimicrobials, *Arch. Dermatol.*, 101, 160, 1969.
63. **Behl, C. R., Mann, K. V., and Bellantone, N. H.,** Transdermal delivery of propranolol. II. Permeation profiles through human skin. Influences of hydration, drug conc., skin sectioning, pH, vehicles, race, sex and anatomical site, Presented at the 131st American Pharmaceutical Association Annual Meeting, Montreal, May 5 — 10, 1984, Basic Pharmaceutics Abstract # 32.
64. **Behl, C. R., Barrett, M., and Flynn, G. L.,** Hydration and percutaneous absorption. II. Influence of hydration on water and alkanol permeation through swiss (furry) mouse skin. Comparison with hairless mouse skin data, *J. Pharm. Sci.*, 70, 1212, 1981.
65. **Wester, R. C. and Maibach, H. I.,** Percutaneous absorption in man and animal: a perspective, in *Cutaneous Toxicity,* Drill, V. and Lazar, P., Eds., Academic Press, New York, 1977, 111.
66. **Bartek, M. J. and La Budde, J. A.,** Percutaneous absorption *in vitro,* in *Animal Models in Dermatology,* Maibach., H. I., Ed., Churchill-Livingstone, New York, 1975, 103.
67. **Campbell, P., Watanabe, T., and Chandrasekharan, S. K.,** Comparison of *in vitro* skin permeability of scopolamine in rat, rabbit and man, *Am. Soc. Exp. Biol.*, 35, 639, 1976.
68. **Bartek, M. J., La Budde, J. A., and Maibach, H. I.,** Skin permeability *in vivo:* comparison in rat, rabbit, pig and man, *J. Invest. Dermatol.*, 58, 114, 1972.
69. **Behl, C. R., El-Sayed, A. A., and Flynn, G. L.,** Hydration and percutaneous absorption. IV. Influence of hydration on n-alkanol permeation through rat skin. Comparison with hairless and swiss (furry) mice, *J. Pharm. Sci.*, 72, 79, 1983.
70. **Behl, C. R., Bellantone, N. H., and Pei, J.,** Effect of the alkyl chain length and anatomical site on the alkanol permeability through fuzzy rat skins, Presented at the 130th Annual Meeting of the American Pharmaceutical Association. New Orleans, LA, April 9, 1983. Basic Pharmaceutics Abstract # 10.
71. **Behl, C. R. and Bellantone, N. H.,** Influence of the alkyl chain length on the *in situ* permeation of n-alkanols through the fuzzy rat skins and comparison with *in vitro* data. Comparison of fuzzy rat and hairless mouse skins results. Presented at the 31st National Meeting of the Academy of Pharmaceutical Sciences, Miami Beach, FL, November 14, 1983. Basic Pharmaceutics Abstract # 38.
72. **Walker, M., Dugard, P. H., and Scott, R. C.,** *In vitro* percutaneous absorption studies: a comparison of human and laboratory species, *Human Toxicol.*, 2, 561, 1983.
73. **DelTerzo, S., Behl, C. R., Nash, R. A., Bellantone, N. H., and Malick, A. W.,** Evaluation of nude rat as a model: effects of short-term freezing and alkyl chain length on the permeabilities of n-alkanols and water. *J. Soc. Cosmet. Chem.*, 37, 297, 1986.
74. **DelTerzo, S.,** Iontophoresis: A Quantitative and Mechanistic Study, Ph. D Thesis, St. John's University, New York, May 1987.
75. **Garcia, B., Marty, J. P., and Wepierre, J.,** Effects of Tween 80 on the permeation rates of alkanols through the hairless rat skin, *Int. J. Pharmaceutics*, 4, 205, 1980.
76. **Sugibayashi, K., Hosoya, K. I., Morimoto, Y., and Higuchi, W. I.,** Effect of absorption enhancer, Azone, on the transport of 5-fluorouracil across hairless rat skin, *J. Pharm. Pharmacol.*, 35, 578, 1985.
77. **Friedlander, S., Dhupar, K. C., Kumar, S., Malick, A. W., and Behl, C. R.,** Skin uptake studies: *in vitro* permeation and retention properties of the acid and salt forms of Ro 23-3544 from various solution formulations using hairless guinea pig skin and finite dose methodology, Presented at the Joint Eastern Regional Meeting of the American Association of Pharmaceutical Scientists. Atlantic City, NJ, June 1988. Abstract # 30.
78. **Char, H., Kumar, S., and Behl, C. R.,** Evaluation of hairless guinea pig as a laboratory animal model for skin permeation studies. Influence of the alkyl chain length on n-alkanol permeability, Unpublished data, Hoffmann-La Roche, Pharmaceutical R&D, Nutley, NJ. 1988.
79. **Anderson, K. E., Maibach, H. I., and Ango, M. D.,** The guinea pig: an animal model for human skin absorption of hydrocortisone, testosterone and benzoic acid?, *Br. J. Dermatol.*, 102, 447, 1980.
80. **Kurihara-Bergstrom, T., Woodworth, M., Feisullin, S., and Beall, P.,** Characterization of the Yucatan miniature pig skin and small intestine for pharmaceutical applications, *Lab. Animal Sci.*, 36, 396, 1986.
81. **Franz, T. J.,** Personal communication, November 1988.
82. **Meyer, W., Schwarz, R., and Neurand, K.,** The skin of domestic mammals as a model for the human skin, with special reference to the domestic pig, in *Current Problems in Dermatology,* Vol. 7 Simmons, G. S., Paster, Z., Klingberg, M. A., and Kaye, M., Karger, Basel, 1978, 39.
83. **McGreesh, A. H.,** Percutaneous toxicity, *Toxicol. Appl. Pharmacol. (Suppl.)*, 2, 20, 1965.
84. **Kydonieus, A. F.,** Transdermal delivery from solid multilayered polymeric reservoir systems, in *Transdermal Delivery of Drugs,* Vol 1, Kydonieus, A. F. and Berner., B., Eds., CRC Press, FL, 1987, chap. 11.
85. **Hunziker, N., Feldman, R., and Maibach, H. I.,** Animal models of percutaneous penetration: comparison in Mexican hairless dogs and man, *Dermatologica*, 156, 79, 1978.
86. **Chukwumerije, O., Nash, R. A., Matias, J., and Orentreich, O.,** Methyl esters of alkyl fatty acids as penetration enhancers for minoxidil percutaneous absorption, Presented at 3rd Annual Meeting of the American Association of Pharmaceutical Scientists, Orlando, FL, October 30 — November 3, 1988. Pharmacetics and Drug Delivery Abstract # 982.

87. **Ritschel, W. A., Nathan, D., and Hussain, A.,** Comparison of skin permeability for two model substances using excised skin from 5 mammalian species, Presented at 3rd Annual Meeting of the American Association of Pharmaceutical Scientists. Orlando, FL, October 30 — November 3, 1988. Pharmacetics and Drug Delivery Abstract # 1001.

88. **Wester, R. C. and Noonan, P. K.,** Relevance of animal models for percutaneous absorption, *Int. J. Pharmaceutics,* 7, 99, 1980.

89. **Feldman, R. J. and Maibach, H. I.,** Percutaneous penetration of steroids in man, *J. Invest. Dermatol.,* 52, 89, 1969.

90. **Bhatt, P. P., Rytting, J. H., Topp., E. M., and Higuchi, T.,** Effect of azone and lauryl alcohol on transport of acetaminophen through shed snake skin, Presented at 3rd Annual Meeting of the American Association of Pharmaceutical Scientists. Orlando, FL, October 30 — November 3, 1988. Pharmacetics and Drug Delivery Abstract # 991.

91. **Brooke, H.,** Hairless mice, *J. Hered.,* 17, 173, 1926.

92. **Crew, F. A. F. and Miskai, L.,** The character hairless in the mouse, *J. Genet.,* 25, 17, 1931.

93. **David, L. T.,** The external expression and comparative dermal histology of hereditary hairlessness in mammals, *Z. Zell. Forsch. Anat.,* 14, 616, 1932.

94. **David, L. T.,** Modification of hair direction and slope of mice and rats, *J. Exp. Zool.,* 68, 519, 1934.

95. **Steinberg, A. G. and Fraser, F. C.,** The expression and interaction of hereditary factors affecting hair growth in mice: external observations, *Canadian J. Res.,* 24, 1, 1946.

96. **Hardy, M. H.,** The histochemistry of the hair follicle in the mouse, *Am. J. Anat.,* 90, 285, 1952.

97. **Chase, H. B. and Montagna, W.,** The development and consequence of hairlessness in the mouse, *Genetics,* 37, 573, 1952.

98. **Montagna, W., Chase, H. B., and Melaragno** The skin of hairless mice. I. The formation of cysts and the distribution of lipids, *J. Invest. Dermatol.,* 19, 83, 1952.

99. **Chase, H. B., Montagna, W., and Malone, D.,** Changes in the skin in relation to the hair growth cycle, *Anat. Rec.,* 116, 75, 1953.

100. **Chase, H. B.,** Growth of the hair, *Physiol. Rev.,* 34, 113, 1954.

101. **Castle, W. E., Dempster, R., and Shurrager, H. C.,** Three new mutations of the rat, *J. Hered.,* 46, 9, 1955.

102. **Winkelmann, R. K.,** The innervation of a hair follicle, *Ann. N.Y. Acad. Sci.,* 83, 400, 1959.

103. **Mann, S. J. and Straile, W. E.,** New observations on hair loss in the hairless mouse, *Anat. Rec.,* 140, 97, 1961.

104. **Yun, J. S. and Montagna, W.,** The skin of hairless mice. III. The distribution of alkaline phosphatase, *Anat. Rec.,* 140, 77, 1961.

105. **Orwin, D. F. G., Chase, H. B., and Silver, A. F.,** Catagen in the hairless mouse, *Am. J. Anat.,* 121, 489, 1967.

106. **Gates, A. H. and Karasek, M. A.,** Hereditary absence of sebaceous glands in the mouse, *Science,* 148, 1471, 1965.

107. **Rugh, R.,** *The Mouse: 1st Reproduction and Development,* Burgess Publishing, 1968, chap. 1.

108. **Gates, A. H., Arundell, F. D., and Karasek, M. A.,** Hereditary defect of the pilosebaceous unit in a new double mutant mouse, *J. Invest. Dermatol.* 52, 115, 1969.

109. **Mann, S. J.,** Hair loss and cyst formation in hairless and rhino mutant mice, *Anat. Rec.,* 170, 485, 1971.

110. **Ferguson, F. G., Irving, G. W., III, and Stedham M. A.,** Three variations of hairlessness associated with albinism in the laboratory rat, *Lab Animal Sci.,* 29, 459, 1979.

111. **Palm, J. and Ferguson, F. G.,** Fuzzy, a hypotrichotic mutant in linkage group I of the Norway rat, *J. Hered.,* 67, 284, 1976.

112. **Bray, G. A.,** The Zucker-fatty rat: a review, *Fed. Proc.,* 36, 148, 1977.

113. **Charles River Laboratories Update,** *New Euthymic Hairless Guinea Pig for Dermatologic Studies,* Vol. 1, Spring 1986. (251 Ballardvale Street, Wilminton, Mass. 01887-0630.)

114. **Garrett, E. R. and Chemburkar, P. B.,** Evaluation, control, and prediction of drug diffusion through polymeric membranes. I. Methods and reproducibility of steady-state diffusion studies, *J. Pharm. Sci.,* 57, 944, 1968.

115. **Garrett, E. R. and Chemburkar, P. B.,** Evaluation, control, and prediction of drug diffusion through polymeric membranes. II. Diffusion of aminophenones through Silastic membranes — a test of the pH partition hypothesis. *J. Pharm. Sci.,* 57, 949, 1968.

116. **Garrett, E. R. and Chemburkar, P. B.,** Evaluation, control and prediction of drug diffusion through polymeric membranes. III. Diffusion of barbiturates, phenylalkylamines, dextromethorphan, progesterone, and other drugs, *J. Pharm. Sci.,* 57, 1401, 1968.

117. **Roseman, T. J. and Higuchi, W. I.,** Release of medroxyprogesterone acetate from a silicone polymer, *J. Pharm. Sci.,* 59, 353, 1970.

118. **Haleblian, J., Runkel, R., Mueller, N., Christopherson, J., and Ng, K.,** Steroid release from silicone elastomer containing excess drug in suspension, *J. Pharm. Sci.,* 60, 541, 1971.

119. **Lovering, E. G. and Black, D. B.,** Diffusion layer effects on permeation of phenylbutazone through polydimethylsiloxane, *J. Pharm. Sci.,* 63, 1399, 1974.
120. **Lovering, E. G., Black, D. B., and Rowe, M. L.,** Drug permeation through membranes. IV. Effect of excipients and various additives on permeation of chlordiazepoxide through polydimethylsiloxane membranes, *J. Pharm. Sci.,* 63, 1224, 1974.
121. **Nakano, M. and Patel, N. K.,** Release, uptake, and permeation behaviour of salicyclic acid in ointment bases, *J. Pharm. Sci.,* 59, 985, 1970.
122. **Flynn, G. L. and Smith, R. W.,** Membrane diffusion. III. Influence of solvent composition and permeant solubility on membrane transport. *J. Pharm. Sci.,* 61, 61, 1972.
123. **Bottari, F., Di Colo, G., Nannipieri, E., Saettone, M. F., and Serafini, M. F.,** Release of drugs from ointment bases. II. *In vitro* release of benzocaine from suspension type aqueous gels, *J. Pharm. Sci.,* 66, 926, 1977.
124. **Kincl, F. A., Benagiano, G., and Angee, I.,** Sustained release hormonal preparations. I. Diffusion of various steroids through polymer membranes, *Steroids,* 11, 673, 1968.
125. **Most, C. F., Jr.,** Filler effects on diffusion in silicone rubber, *J. Appl. Polymer Sci.,* 14, 1019, 1970.
126. **Bellantone, N. H. and Behl, C. R.,** Azone® (dodecylazacycloheptan-2-one) altered skin permeability of model compounds. II, Presented at the 35th national meeting of the Academy of Pharmaceutical Sciences, Miami Beach, FL, November 13 — 17, 1983. Basic Pharmaceutics Abstract # 68.
127. **Behl, C. R., Pei, J. Y., and Malick, A. W.,** Skin Permeability profile of 14C diazepam, Presented at the 133rd annual meeting of the American Pharmaceutical Association, San Francisco, CA, March 16 — 20, 1986. Pharmacodynamics and Drug Disposition Abstract # S 814-2.
128. **Behl, C. R., Malick, A. W., and Goldberg, A. H.,** Transdermal delivery of Ro 22-1327. I. Partitioning and *in vitro* skin permeability studies. Correlation between human and rat skins, and Silastic® permeability studies, Presented at the 39th National Meeting of the Academy of Pharmaceutical Sciences. Minneapolis, MN, October 20 — 24, 1985. Basic Pharmaceutics Abstract # 25.
129. **Kumar, S., Dhupar, K. C., Patel, S. B., Weinrib, A. B., and Behl, C. R.,** *In vitro* skin uptake studies with retinoids using hairless mouse skin: determination of permeation rates, correlation of skin to Silastic® data and flux vs. solubility considerations, Presented at the First Annual Eastern Regional Meeting of the American Association of Pharmaceutical Scientists. Atlantic City, NJ, September 13 — 14, 1987, Pharmaceutics and Drug Delivery Abstract # 18.
130. **Flynn, G. L., Behl, C. R., Kurihara, T., Smith, W. M., Fox, J. L., Durrheim, H. H., and Higuchi, W. I.,** Correlation and prediction of mass transport across membranes. III. Boundary layer and membrane resistances to diffusion of water and n-alkanols through polydimethylsiloxane (Silastic) membranes, Presented at the 26th National Meeting of the Academy of Pharmaceutical Sciences, Anaheim, CA, April 21 — 26, 1979. Basic Pharmaceutics Abstract # 30.
131. **Gary-Bobo, C. M., Di Polo, R., and Solomon, A. K.,** Role of hydrogen-bonding in nonelectrolyte diffusion through dense artificial membranes, *J. Gen. Physiol.,* 55, 369, 1969.
132. **Di Polo, R., Sha'afi, R. I., and Solomon, A. K.,** Transport parameter in a porous cellulose acetate membrane. *J. Gen. Physiol.,* 55, 63, 1970.
133. **Barry, B. W. and Brace, A. R.,** Permeation of oestrone, oestradiol, oestriol and dexamethasone across cellulose acetate membrane, *J. Pharm. Pharmacol.,* 29, 397, 1977.
134. **Barry, B. W. and El Eini, D. I.,** Influence of non-ionic surfactants on permeation of hydrocortisone, dexamethasone, testosterone and progesterone across cellulose acetate membrane, *J. Pharm. Pharmacol.,* 28, 219, 1976.
135. **Shah, V. P., Elkins, J. S., Hanus, J. P., and Skelly, J. P.,** Automated method for *in vitro* release of hydrocortisone (HC) from creams, ointments and lotions, Presented at the Third Annual Meeting of the American Association of Pharmaceutical Scientists, Orlando, FL, October 1988. Abstract # PDD—883.
136. **Srinivasan, V., Sims, S. M., Higuchi, W. I., Malick, A. W., Behl, C. R., and Pons, S.,** Iontophoretic transport of drugs: a constant voltage approach, in *Modulated Controlled Release Systems,* CRC Press, Boca Raton, FL, in press.
137. **Masada, T., Higuchi, W. I., Rohr, U., Fox, J., Malick, A. W., Pons, S., Goldberg, A. H., and Behl, C. R.,** Examination of iontophoretic transport of ionic drugs across the skin. I. Baseline studies with the four electrode system, *Int. J. Pharmaceutics,* in press.
138. **Higuchi, W. I.,** Iontophoretic delivery of peptides and proteins: *in vitro* studies and mechanistic analysis, Presented at the International Conference on Pharmaceutical Sciences and Clinical Pharmacology, Jerusalem, Israel, May 29 — June 3, 1988.
139. **Hunke, W. A. and Matheson, L. E., Jr.,** Mass transport properties of Co (polyether) polyurethane membrane. II. Permeability and sorption characteristics, *J. Pharm. Sci.,* 70, 1313, 1981.
140. **Behl, C. R., Bellantone, N. H., Dethlefsen, L., and Goldberg, A. H.,** Design of transdermal delivery systems. I. Characterization of permeability behaviors of rate controlling synthetic membranes. Evaluation of the alkyl chain length and vehicle influences, Presented at the 131st Annual Meeting of the American Pharmaceutical Association, Montreal, May 5 — 10, 1984. Basic Pharmaceutics Abstract # 1.

141. **Radwan, M. A., Goldman, D., Behl, C. R., Kumar, S., and Friedlander, S. L.,** Development of *in vitro* release method for semisolid products, Presented at the 3rd Annual Meeting of the American Association of Pharmaceutical Scientists, Orlando, FL, October 30 — November 3, 1988. Abstract # PDD—885.

142. **Dyer, A., Hayes, G. G., Wilson, J. G., and Catterall, R.,** Diffusion through skin and model systems, *Int. J. Cosmet. Sci.,* 1, 91, 1979.

143. **Nacht, S. and Yeung, D.,** Artificial membranes and skin permeability, in *Percutaneous Absorption,* Bronaugh, R. L. and Maibach, H. I., Eds., Marcel Dekker, New York, 1985, chap. 29.

144. **O'Neill, W. P.,** Membrane systems, in *Controlled Release Technologies: Methods, Theory, Applications,* Vol. 1, Kydonieus, A. F., Ed., CRC Press, Boca Raton FL, 1980, 129.

145. **Nakano, M., Kuchiki, A., and Arita, T.,** The effect of charges on permeabilities of drugs through collagen membranes, *Chem. Pharm. Bull. (Tokyo),* 24, 2345, 1976.

146. **Washitake, M., Takashima, Y., Tanaka, S., Anmo, T., and Tanaka, I.,** Drug permeation through egg shell membranes, *Chem. Pharm. Bull., (Tokyo),* 28, 2855, 1980.

147. **Chien, Y. W.,** Microsealed drug delivery systems: theoretical aspects and biomedical assessments, in *International Symposium on Recent Advances in Drug Delivery Systems,* Anderson, J. M. and Kim, S. W., Eds., Plenum Press, New York, 1984, 367.

148. **Albery, W. J., Couper, A. M., Hadgraft, J., and Ryan, C.,** Transport and kinetics in two phase systems, *J. Chem. Soc. (Faraday Trans. 1),* 70, 1124, 1974.

149. **Albery, W. J., Burke, J. F., Leffler, E. B., and Hadgraft, J.,** Interfacial transfer studied with a rotating diffusion cell, *J. Chem.Soc. (Faraday Trans. 1),* 72, 1618, 1976.

150. **Albery, W. J. and Hadgraft, J.,** Percutaneous absorption: interfacial transfer kinetics, *J. Pharm. Pharmacol.,* 31, 65, 1979.

151. **Guy, R. H. and Fleming, R.,** The estimation of diffusion coefficients using the rotating diffusion cell, *Int. J. Pharmaceutics,* 3, 143, 1979.

152. **Tanaka, M., Fukuda, H., and Nagai, T.,** Permeation of drug through a model membrane consisting of Millipore filter with oil, *Chem. Pharm. Bull. (Tokyo),* 26, 9, 1978.

153. **Shah, V. P.,** Evaluation of transdermal drug delivery systems, Presented at the Workshop on Transdermal Delivery of Drugs, Bethesda, MD, May 23 — 24, 1988.

154. **Behl, C. R., Barrett, M., and Flynn, G. L.,** Percutaneous absorption of water and n-alkanols through skin of the mature Swiss (furry) mouse: comparison with the human and the hairless mouse skin behavior, Presented at the 127th Annual Meeting of the American Pharmaceutical Association, Washington, D.C., April 19 — 24, 1980. Basic Pharmaceutics Abstract # 35.

155. **Desai, K. J.,** Biophysical Approach to the Study of Intestinal Transport of Drugs, Ph.D. thesis, The University of Michigan, Ann Arbor, MI, 1976.

156. **Hwang, S.,** Systems Approach to the Study of Vaginal Drug Absorption in the Rabbit, Ph.D. thesis, The University of Michigan, Ann Arbor, MI, 1976.

157. **Behl, C. R.,** Systems Approach to the Study of Vaginal Drug Absorption in the Rhesus Monkey, Ph.D. thesis, The University of Michigan, Ann Arbor, MI, 1979.

158. **Idson, B. H. and Behl, C. R.,** Drug structure vs. skin permeation, in *Transdermal Delivery of Drugs,* Vol. 3, Kydonieus, A. F. and Berner, B., Eds., CRC Press, Boca Raton FL, chapter 4, 1987.

159. **Bond, J. R. and Barry, B. W.,** Hairless mouse skin is limited as a model for assessing the effects of penetration enhancers in human skin, *J. Invest. Dermatol.,* 90, 810, 1988.

160. **Sims, S. M., Higuchi, W. I., and Srinivasan, V.,** A quantitative assessment of the parallel lipid-aqueous pore pathway model for the iontophoretic transport of weak electrolytes, Presented at the 3rd Annual Meeting of the American Association of Pharmaceutical Scientists, Orlando, FL, October 30 — November 3, 1988. Pharmaceutics and Drug Delivery Abstract # PDD 1035

Chapter 2

PENETRANT CHARACTERISTICS INFLUENCING SKIN ABSORPTION

Kelly L. Smith

TABLE OF CONTENTS

I. INTRODUCTION

Absorption or transport of drugs, toxicants, or other chemicals into or through the skin depends on a number of factors: characteristics of the penetrant, condition and type of skin, other chemicals (e.g., vehicles or enhancers) present with the penetrant, and external conditions such as temperature, humidity, and occlusion. Under most conditions, the factor with perhaps the greatest influence on the rate or extent of skin absorption is the character of the penetrant, and that is the focus of this chapter.

Chemicals are transported into and through the skin by a solution-diffusion process. The penetrant must dissolve in the skin, diffuse across the skin, and partition into the body fluids or tissues beneath the skin. Due to the extraordinary barrier that skin represents to most penetrants, diffusion across the skin is typically the slowest, and therefore rate-controlling, step in this process. In this case, the rate of transport across the skin can be described by the following approximation of Fick's first law:

$$J = \frac{D\Delta C_m}{\ell} \tag{1}$$

where J is the flux of penetrant across the skin (typically in units of $g/cm^2/s$, D is the diffusivity of penetrant in the skin in cm^2/s, ΔC_m is the difference in penetrant concentration within the skin between one side of the skin and the other, in units of g/cm^3, and ℓ is the thickness of the skin in centimeters. It is apparent from Equation 1 that the flux is constant if the diffusivity and concentration difference are constant; conditions that control this are discussed in the following sections. Equation 1 is sometimes written in a form that includes a partition coefficient (k); in that case, the concentration difference is that between the phases on each side of the skin, rather than within the skin. In order to be accurate, however, a different partition coefficient is usually required for each side of the membrane.

Equation 1 in this simple form is valid only (1) for homogenous materials, (2) where the concentration difference is constant with time, and (3) where the diffusivity is not dependent on concentration; frequently, none of these are the case in transport across the skin. For example, the skin is generally recognized as a nonhomogeneous structure, in which the properties of the stratum corneum, near the surface, are very different from those of the underlying epidermis and dermis.[1,2] For this reason, one often refers to an effective diffusivity and an effective skin thickness in order to simplify calculations. In Equation 1, both the diffusivity and the concentration difference are closely tied to properties of the penetrant, as we will see in subsequent sections of this chapter. However, it is important to note that both of these properties are also dependent on other factors, including vehicles. A lack of understanding of this dependence, and the fundamentals of transport across skin, have resulted in many misleading or confusing results published in recent years.

It has been shown that in most cases transport across the skin is controlled by transport across the stratum corneum.[3,4] For this reason, absorption into the bulk of the skin (i.e., beneath the stratum corneum) and transport through the skin are both dependent on the same factors, and they occur at approximately the same rates. Thus, although Equation 1 is written in terms that describe transport across the skin, it can be used to describe absorption into the skin as well.

The following sections describe first the chemical and physical properties that influence skin absorption and transport, and subsequently the influence that these factors have on experimental design. The last section describes methods of determining the principal penetrant characteristics of interest.

II. CHEMICAL PROPERTIES

Various chemical characterisitcs, including solubility, lipophilicity, ionization, and stability, are important in influencing transport of penetrants across skin. Referring to Equation 1, these chemical properties of the penetrant greatly control the concentration difference within the membrane, C_m, but generally have little effect on the diffusivity. In most cases, diffusivity is more closely related to physical properties of the penetrant.

A. Solubility and Partitioning

The solubility of the penetrant in the various phases present in the skin and its surroundings plays a large part in determining the rate of penetration. In a typical case, the penetrant will be present on the skin surface either dissolved in or dispersed in a vehicle of some sort. While it is the concentration of penetrant within the skin that controls the rate of transport, that concentration is dependent on the concentration and solubility (i.e., thermodynamic activity) of penetrant in the vehicle on the skin surface. A critical point in understanding skin transport is that penetrant concentration in the vehicle does not determine the rate of transport. Rather, the thermodynamic activity (or more rigorously, the chemical potential) of the penetrant in the vehicle determines the rate of transport. This can perhaps best be visualized by referring to a sketch of a cross-section of the skin, showing concentrations and thermodynamic activities of the penetrant at steady state (Figure 1). In Figure 1A, the penetrant is dissolved in a vehicle in which it is highly soluble; its concentration is high, but its activity is less than unity. The thermodynamic activity of the penetrant is identical within the membrane at its interface with the vehicle, so the concentration of penetrant within the membrane at that interface is less than saturation, and the rate of transport is less than maximum. In Figure 1B, the penetrant is dissolved, with excess penetrant present, in a vehicle in which it is only moderately soluble. In this case, its concentration in the vehicle is less than the case in Figure 1A, but its thermodynamic activity is unity, and the corresponding concentration within the membrane is higher, resulting in a higher rate of transport than was the case in Figure 1A. All other things constant, the most rapid rate of transport will occur when the penetrant is in pure or saturated form (i.e., its thermodynamic activity is unity) on the skin surface. If the penetrant is dissolved in a vehicle at a concentration lower than saturation, its activity is less than unity, and the concentration within the membrane will be less than saturation, resulting in a lower flux.

It should be noted that measurement of the concentration of penetrant within the skin is not straightforward, and it is typical to measure the much more easily determined concentration in the vehicle. Without determining the solubility (and therefore the activity) in the vehicle, however, results of vehicle studies can be misleading. For example, there have been numerous studies comparing transdermal fluxes of penetrants from vehicles containing a fixed concentration of penetrant. While such studies can be informative, they are typically more a measure of thermodynamic activity of the penetrant in the vehicle than of any flux-enhancing effects.

The discussion so far has focused on the phase boundary between the vehicle and the stratum corneum. While this is perhaps the most important phase boundary, it should be pointed out that once the penetrant has crossed the stratum corneum, it must partition into the underlying layers of epidermis, dermis, and circulatory system. These tissues are typically more hydrophilic than is the stratum corneum and can present a barrier to transport of extremely hydrophobic penetrants.

The stratum corneum is generally modeled as consisting of a continuous lipid phase surrounding a large number of thin parallel protein "plates" (the brick-and-mortar model).[5,6] If this structure remains intact, penetration occurs via one of two routes: either by diffusion through both the hydrophobic lipid phase and the hydrophilic protein phases, or (for lipophilic

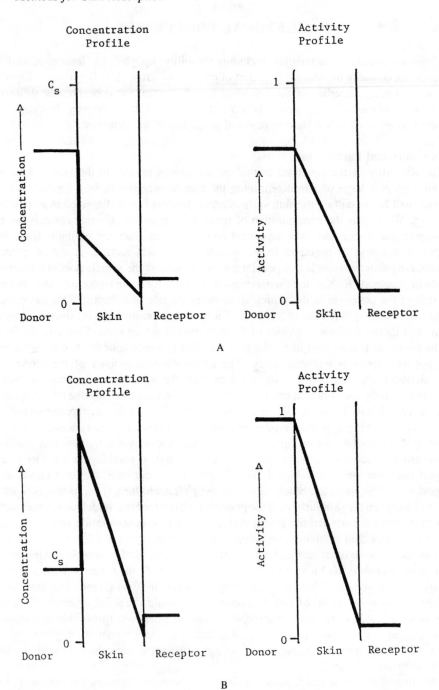

FIGURE 1.   (A) Concentration and thermodynamic activity profiles of penetrant during transdermal permeation. Penetrant at high concentration in donor, but less than saturation. (B) Concentration and thermodynamic activity profiles of penetrant during transdermal permeation. Penetrant at low concentration in donor, but saturated.

penetrants) by diffusion via a tortuous pathway solely through the lipid phase. A penetrant with extremely low solubility in either the hydrophobic or hydrophilic phases will be essentially blocked by the phase, and will generally have a low transdermal flux. Conversely, penetrants with high solubilities in both phases will exhibit high fluxes, or at least will not be limited by solubility considerations.

As can be seen from the sketches in Figure 1, thermodynamic activity is continuous across phase boundaries, while concentration is not. The discontinuities in concentration across phase boundaries are described by partition coefficients. A partition coefficient of one indicates no discontinuity, which results when the solubility of the penetrant is identical in each phase. The partition coefficient is usually assumed to be concentration independent, and can be approximated by the ratio of concentrations or solubilities in each phase.

The principal use of partition coefficients is in predicting relative rates of transport or absorption from relatively quick and easy partitioning experiments between two liquid phases. Perhaps the most predictive results have been obtained from octanol/water and from mineral oil/water partition coefficients. Lien and Tong[7] have shown an increasing flux as a function of increasing octanol/water partition coefficient over a very wide range for two different homologous series of compounds: alcohols and steroids. It should be noted however, that the fluxes of alcohols were about two to three orders of magnitude higher than those of steroids at equal partition coefficients, so partition coefficient alone does not tell the entire story. Similarly, Michaels, et al.[5] found good correlation of transdermal flux with mineral oil/water partition coefficient for drugs with a wide range of properties over about five orders or magnitude in partition coefficient. Other studies with various alkanes, alcohols, ethers, oils, and other solvents have demonstrated some correlations with transdermal flux, but typically over small ranges of partition coefficient.[8-11] The partition coefficient by itself is not usually a good indicator of solubility in the membrane, because it does not predict the absolute solubility in the membrane. However, the partition coefficient combined with solubility measurements can be a good indicator of solubility in the membrane. Nevertheless, it does not predict diffusivity, which can also play an important role in skin absorption, as discussed in a later section.

B. IONIZATION

If the penetrant is ionizable, both charged and uncharged species are present, in quantities dependent on pH. Generally, transport of ionized species occurs much less rapidly than transport of the base or unionized species. For example, the transdermal flux of scopolamine was shown to be substantially higher at pHs above the pKa of the weak base.[12] While it is possible that facilitated transport of ion pairs via a carrier vehicle may result in enhanced transdermal flux,[13] in general higher fluxes will be obtained by maintaining the pH such that the penetrant is unionized. It should be noted that nonphysiological pHs may also change properties of the skin that could affect solubility, partitioning, or binding, resulting in changes in transdermal penetration. The use of an electric current to enhance transport of ionized species (iontophoresis) has been shown to be feasible,[14-16] but is beyond the scope of this chapter.

C. STABILITY

As we have seen, the driving force for absorption or transport of any penetrant is proportional to the concentration gradient of that penetrant within the skin. It follows that any degradation or other modification of the penetrant to another species reduces the concentration of the penetrant species in the skin, and therefore reduces the flux. Further, the concentration of the product(s) that is formed increases in the skin, and it is transported more rapidly across the skin. This applies not only to degradation that occurs on the skin surface or in the vehicle, but also to degradation within the skin. It is now widely recognized that the skin is a viable organ that actively metabolizes chemicals that enter it. Certainly that is not always the case, but it is not unusual for metabolism of penetrants to occur. This is not usually a problem for *in vitro* experiments with skin or skin models, but should be anticipated for *in vivo* studies. Effects of metabolism may be studied *in vitro* if the viability of skin samples is maintained. Degradation due to oxidation, light, or other factors must be considered for both *in vitro* and *in vivo* experiments.

III. PHYSICAL PROPERTIES

The principal physical properties that affect skin absorption or transport are molecular weight and molecular size. These factors directly influence the diffusivity, on which the rate of transport depends, as seen in Equation 1. Diffusivity is a kinetic term, and is a rough measure of the ease with which a molecule can move about within a medium (in this case, the skin). The larger the molecule, the more difficult it is to move about, and the lower the diffusivity.

For small molecules in homogeneous media, the diffusivity is roughly proportional to the inverse of the cube root of the molecular volume (or molecular weight), which is not a very strong dependence. For larger molecules (and especially for large, nonspherical molecules) in nonhomogeneous media (such as the skin), the dependence of diffusivity on molecular weight is more striking. This is essentially due to steric hindrance of motion as the molecular size approaches the size of the available pathways for diffusion. Thus, diffusivities through crystalline polymers, where diffusing species must move around crystalline regions, can be several orders of magnitude lower than diffusivities through amorphous polymers, where diffusion can occur in more or less a straight line.[17]

It is interesting to note that, although solubility can be limiting, most penetration enhancers increase the diffusivity of the skin more than they increase the solubility of penetrants within the skin.[5,9] (The relative effects of enhancer on solubility and diffusivity can be determined by measuring these properties, as discussed in Section V.B) The increase in Occlusion of the skin also increases diffusivity via a swelling process (hydration).

IV. PENETRANT INFLUENCE ON EXPERIMENTAL DESIGN

Most measurements of skin absorption are made using two-compartment diffusion cells separated by a layer of skin or other membrane. Typically, the penetrant is placed on one side of the skin, and its rate of transport to the other side is measured. Two distinct types of diffusion cells are used: those with a reservoir of stirred liquid on each side and those with liquid on one side and a gas-containing (or no) chamber on the other. The latter (often called Franz cells) are often used with just one (receptor) chamber, and the penetrant is placed directly onto the skin as a liquid or ointment.[18] The two-liquid-compartment type of cell typically results in more skin hydration than does the Franz cell, which can increase the diffusivity of penetrants in the skin (and decrease the time lag); it can also affect solubility. Both of these types of cells are addressed in this section.

A key influence that penetrant properties have on experimental design is on the time scale for measurement of skin absorption. When penetrant is placed in contact with skin, there is often a considerable period of time before the steady-state concentrations depicted in Figure 1 are achieved (it should be noted that steady state is achieved only if the penetrant concentration in the donor chamber remains constant). Barring irreversible binding of the penetrant by the skin, this time delay is dependent almost entirely on the diffusivity of penetrant in the stratum corneum. As discussed in Section V, the time lag, as the characteristic time delay is termed, is also highly dependent on thickness, but for a given type of skin, the thickness can be considered constant. Thus, penetrants with higher molecular weights will result in longer time lags than will lower-molecular-weight penetrants. Steady-state fluxes are generally achieved within about three time lags (not one).

Careful control of temperature on both sides of the skin is imperative in skin absorption experiments. Solubility and diffusivity are both activated properties, and increase rapidly with increasing temperature. Often, studies are designed to maintain a normal skin temperature of about 32°C, but variations of even one degree can cause significant differences in rates of absorption.

A. VEHICLE

The term "vehicle" is usually used to describe a liquid or semisolid in which the penetrant is dissolved or dispersed. Thus, vehicle effects are more relevant for the Franz-cell type of apparatus than for the two-liquid-compartment diffusion cell.

As discussed above, a vehicle present with the penetrant can influence the concentration of penetrant within the membrane, as well as its diffusivity in the skin, and therefore its rate of transport. Barring any effects of the vehicle on the skin or membrane, however, the rate of transport of penetrant from a saturated solution of penetrant in any vehicle will be identical. This is because the thermodynamic activity of penetrant in any saturated solution is unity, and this results in an identical concentration (saturation) in the membrane regardless of the concentration in the vehicle. In tests with skin, however, it is perhaps more common for a vehicle to affect the permeability of the skin than not. For example, water hydrates the skin, swelling the tissue and increasing its diffusivity (and solubility of hydrophilic penetrants). Various penetration enhancers penetrate the skin themselves, disrupting the skin structure or fluidizing its lipids, resulting in increased rates of absorption.[19-21] Finally, there are some vehicles, such as polyethylene glycol,[22] that appear to slow the rate of absorption, although the mechanism is not well understood.

It should also be noted that vehicle effects are not instantaneous; that is, the same types of time-lag phenomena that occur with penetrants occur with vehicles as well, since they are in fact penetrants themselves. Since vehicles and other penetrants do not have the same diffusivity, the time lags for different components will be different. Further complicating matters, the time lags will not be independent of each other unless neither affects the membrane or skin. As discussed in Section V, absorption should be measured as a function of time until steady state is reached; that is, empirical evidence is more reliable than theoretical calculations in this case.

B. DONOR CHAMBER

The donor chamber is meant to comprise the environment on the penetrant side of the skin or membrane. In the case of Franz cells, that includes the penetrant, vehicle, and surrounding gas; in the case of two-liquid-compartment cells, that includes whatever liquid is present in the donor chamber. The considerations are generally different in these cases and will be discussed separately.

In Franz cells, one of the important considerations is volatility of the penetrant or of any vehicle present. This can result in at least three effects on skin penetration. First, evaporation of one or more components can change the composition of the penetrant mixture at the skin surface, which can change the thermodynamic activity. Second, evaporation reduces the quantity of penetrant or vehicle available to penetrate, which can make attainment of steady state difficult or impossible. Third, in the case of very volatile components, a significant amount of evaporation could lower the temperature of the skin surface, which would lower the rate of transport. A similar consideration with Franz cells is evaporation of water from the skin, resulting in dehydration and lowering its diffusivity. The evaporation rate of penetrant, vehicle, or water is dependent on the rate of air flow across the skin surface. In some studies, this should be carefully controlled, and the quantities of penetrant and/or vehicle that evaporate should be measured; this is the case with fragrances, insect repellents, and the like. In other studies, such as with a solid drug dispersed in a nonvolatile ointment, evaporation is generally not a concern.

A final concern with Franz cells in some cases is delivery of the penetrant to the skin surface. If the penetrant is dissolved or dispersed in a nonvolatile vehicle, it must diffuse through that vehicle to the skin surface before it can be absorbed. In almost all cases, this process is rapid compared with diffusion through the skin, and is not therefore a problem. However, in cases where the penetrant is exceedingly insoluble in the vehicle, there may be a diffusion boundary layer within the vehicle that slows the rate of absorption. In this type of cell, there is not much that can be done to avoid this, other than changing vehicles to increase the solubility of the penetrant.

With the two-liquid-compartment cells, volatility is not usually a concern, other than to maintain sufficient liquid volume in each compartment to fully cover the skin. There are two other key differences in the donor chamber of this type of cell from Franz cells. First, the liquid in the donor chamber almost always affects the skin in some way. Typically, the liquid is primarily water, which results in hydration of the skin, increasing the rate of transport of most penetrants. The second difference is the capability of stirring the liquid. Unless the donor chamber is filled with a pure liquid penetrant, it must be stirred to avoid a diffusion boundary layer near the skin surface. In most cases, the flux as a function of stirring rate should be measured to ensure that sufficient stirring is maintained to avoid boundary layers. The most reliable absorption data is usually obtained from well-stirred saturated solutions of penetrant containing excess penetrant.

C. RECEPTOR CHAMBER

In both Franz cells and two-liquid-compartment cells, the receptor chamber typically contains an aqueous liquid. Often, this is physiological saline, Ringer's solution, or some other physiologically significant solution. It is generally thought that this solution does not overly hydrate the skin, since it is exposed to a highly aqueous environment *in vivo* as well. The two key factors of concern in the receptor chamber, other than temperature, are solubility and stirring. Both of these are important in allowing the penetrant to be transported away from the skin after it has passed through, avoiding a concentration build-up within the skin, which would slow transport. Stirring in the receptor chamber is usually not as critical as it is in the donor chamber, but it cannot be ignored. The concentration of penetrant in the receptor solution should be maintained low (typically less than 10%) compared with its solubility in the solution to avoid concentration build-up within the membrane. If this is difficult to do, for example with highly insoluble penetrants, solubility-enhancing components can be added to the receptor fluid. Alcohols and glycols have been most commonly used for this purpose,[23] but their potential effects on the skin, as discussed in Section IV.A, must be considered.

V. DETERMINATION OF PENETRANT CHARACTERISTICS

There are several properties of penetrants discussed above that influence rates of skin absorption. These include structure, molecular weight, pKa, volatility, solubility, partition coefficients, and diffusivity. These properties can be determined in basically three methods: searching the literature, theoretical calculations, and experimental measurements. These are now discussed briefly.

A. LITERATURE

It is assumed that chemical structure, molecular weight, and pKa of a penetrant of interest are readily available, and determination of these properties will therefore not be discussed. Volatility is a property that is not usually available in the literature, and is best determined experimentally. An exception is the category of organic solvents, for which volatilities are often available in product literature. Similarly, diffusivities are rarely reported in the literature, and are best determined experimentally.

The best help literature searching can offer is in the area of solubility and partition coefficients. As discussed in the previous section on solubility, the concentration of penetrant relative to its solubility at the skin or membrane surface is a critical parameter to understanding transdermal absorption. Solubilities of commercially available drugs, pesticides, solvents, etc. in water and other common solvents are frequently reported in product literature, various references,[24-26] and published articles.

Most of the literature on transdermal penetration reports partition coefficients, rather than solubilities. As described earlier, partition coefficients can be valuable predictors of transdermal absorption, when combined with solubilities and diffusivities. Nearly all partition coefficients are measured between water and an organic solvent such as octanol, mineral oil, or hexane.[27] Some researchers, however, have also measured partition coefficients between a vehicle and either skin or stratum corneum.[9] Representative listings of partition coefficients for various drugs are given in References 7 to 11.

B. EXPERIMENTAL MEASUREMENTS AND CALCULATIONS

The principal factors that are important to skin absorption and that can be easily measured or calculated are volatility, solubility, partition coefficients, and diffusivity.

Volatility is a measure of the evaporation rate of a compound, and is highly dependent on temperature. The temperature during any volatility measurement must be noted and controlled. Volatility is also directly proportional to surface area, and should be normalized to this. Finally, the volatility of any substance is dependent on other components mixed with it. The volatility is measured most easily by simply monitoring weight loss from a sample with known surface area as a function of time. The rate of weight loss, in for example $\mu g/cm^2/sec$, can be compared with the expected or measured transdermal flux to determine if it is significant. Alternatively, measurement of volatility can be made via collection of the vapor evaporated from a surface and analysis by gas or liquid chromatography.

The solubility of a penetrant in a solvent is best measured by allowing excess penetrant to equilibrate in the solvent while stirring at a constant temperature over several days. It is important to realize that solubility is an equilibrium property, not a kinetic one, and sufficient time must be allowed for maximum dissolution to occur. After equilibration, a sample of the liquid is filtered, diluted with solvent, and analyzed by ultraviolet spectrophotometry, gas or liquid chromatography, or other means. To avoid precipitation of penetrant from the saturated solution in the sample, the filter is typically heated to the same temperature as the solution. The solution can be stored in a constant-temperature water bath with shaking or magnetic stirring. Care must be taken to avoid including any undissolved particles in the measurement. An alternative to using filtration is to use a two-compartment diffusion cell with excess penetrant in solvent on one side, and pure solvent on the other side, separated by a glass frit or other microporous membrane. The frit or membrane allows penetrant to diffuse into the solvent until it reaches equilibrium without allowing solid particles to reach the solvent side. Solubility of penetrants in skin can also be measured by equilibrating a skin sample with the penetrant or penetrant dissolved in a vehicle and subsequently extracting from the skin sample the penetrant that has dissolved in it.

Partition coefficients can be calculated from solubility measurements, simply as the ratio of the solubility in one solvent to that in another solvent (typically water). Alternatively, they can be measured via liquid-liquid extraction. In this case, a fixed quantity of penetrant is dissolved in one liquid, and this solution is then shaken with the other liquid in a separatory funnel at a constant temperature for at least 24 hours. The ratio of concentrations of penetrant in the two liquids at equilibrium is the partition coefficient.

In contrast with solubility, which is an equilibrium property, diffusivity is a kinetic property, and is therefore measured or calculated from kinetic studies. There are two common ways of determining diffusivity of a penetrant in skin or another membrane. The first method involves simply measuring the transdermal flux at early times until a steady-state flux is reached. A representative plot of the concentration of penetrant in the receptor fluid in such an experiment is shown in Figure 2. The time lag before steady state is reached is characteristic of the diffusivity of the penetrant in the membrane, and can be used to calculate the diffusivity.

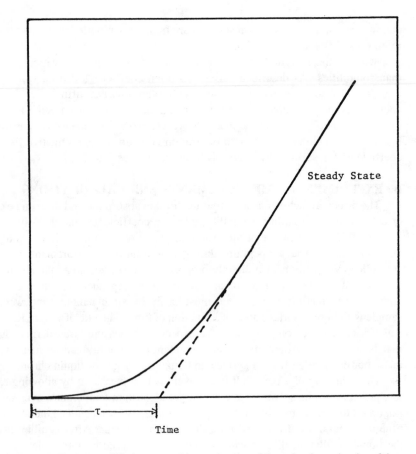

FIGURE 2.   Cumulative mass of penetrant diffusing across skin as a function of time, showing estimation of time lag.

The time lag, τ, is the time obtained from extrapolating the steady-state portion of the curve back to the abscissa, and is defined by the following equation:

$$\tau = \frac{\ell^2}{6D} \tag{2}$$

where ℓ is the thickness of the membrane in cm (in the case of skin, ℓ is usually taken to be the thickness of the stratum corneum), and D is the diffusivity in cm²/s. As seen from Figure 2, steady state is essentially achieved within about three time lags. The other method for determining diffusivity involves measuring the rate of either penetrant uptake by or desorption from a membrane or skin sample. Typically, one measures the rate of desorption of penetrant from a sample that has been equilibrated with a saturated solution of penetrant in water or other vehicle. The diffusivity can be calculated from the following equation:

$$\frac{M_t}{M_o} = 4\left(\frac{Dt}{\pi\ell^2}\right)^{1/2} \tag{3}$$

for desorption of the first 60% of the dissolved penetrant, or

$$\frac{M_t}{M_o} = 1 - \frac{8}{\pi^2}\exp\left(\frac{-\pi^2 Dt}{\ell^2}\right) \tag{4}$$

for desorption of the last 60% of the dissolved penetrant, where M_t is the mass of penetrant desorbed at any time, M_o is the total mass of penetrant absorbed, τ is time, and ℓ is the skin thickness.[28] Calculation of diffusivity from Equations 3 or 4 is straightforward and can be done in conjunction with measuring penetrant solubility in the skin.

REFERENCES

1. **Scheuplein, R. J. and Blank, I. H.,** Permeability of the skin, *Physiol. Rev.,* 51, 702, 1971.
2. **Montagna, W. and Parakkal, P. F.,** *The Structure and Function of Skin,* Academic Press, New York, 1974.
3. **Scheuplein, R. J.,** Mechanism of percutaneous absorption. I. Routes of penetration and the influence of solubility, *J. Invest. Dermatol.,* 45, 334, 1965.
4. **Flynn, G. L., Durrheim, H. H., and Higuchi, W. I.,** Permeation of hairless mouse skin. II. Membrane sectioning techniques and influence on alkanol permeabilities, *J. Pharm. Sci.,* 70, 52, 1980.
5. **Michaels, A. S., Chandrasekaran, S. K., and Shaw, J. E.,** Drug permeation through human skin: theory and in vitro experimental measurement, *A. I. Ch. E. J.,* 21, 985, 1975.
6. **Berner, B. and Cooper, E. R.,** Models of skin permeability, in *Transdermal Delivery of Drugs,* Vol. 2, Kydonieus, A. F. and Berner, B., Eds., CRC Press, Boca Raton, FL, 1987, 41.
7. **Lien, E. J. and Tong, G. L.,** Physiochemical properties and percutaneous absorption of drugs, *J. Soc. Cosmet. Chem.,* 24, 371, 1973.
8. **Anjo, D. M., Feldmann, R. J., and Maibach, H. I.,** Methods of predicting percutaneous penetration in man, in *Percutaneous Penetration of Steroids,* Mauvais-Jarvais, P., Wepierre, J., and Vickers, C. F. H., Eds., Academic Press, New York, 1980, 31.
9. **Southwell, D. and Barry, B. W.,** Penetration enhancers for human skin: mode of action of 2-pyrrolidone and dimethylformamide on partition and diffusion of model compounds water, n-alcohols, and caffeine, *J. Invest. Dermatol.,* 80, 507, 1983.
10. **Mollgaard, B. and Hoelgaard, A.,** Vehicle effect on topical drug delivery. I. Influence of glycols and drug concentration on skin transport, *Acta Phar. Suec.,* 20, 433, 1983.
11. **Bronaugh, R. L. and Congdon, E. R.,** Percutaneous absorption of hair dyes: correlation with partition coefficients, *J. Invest. Dermatol.,* 83, 124, 1984.
12. **Chandrasekaran, S. K., Michaels, A. S., Campbell, P. S., and Shaw, J. E.,** Scopolamine permeation through human skin in vitro, *A. I. Ch. E. J.,* 22, 828, 1976.
13. **Barker, N., Hadgraft, J., and Wotton, P. K.,** Facilitated transport across liquid/liquid interfaces and its relevance to drug diffusion across biological membranes, *Farad. Discuss. Chem. Soc.,* 77, 97, 1984.
14. **Gadsby, P. E.,** Visualization of the barrier layer through iontophoresis of ferric ions, *Med. Instr.,* 13, 281, 1979.
15. **Burnette, R. and Ongpipattanakul, B.,** Characterization of the permselective properties of excised human skin during iontophoresis, *J. Pharm. Sci.,* 76, 765, 1987.
16. **Rolf, D.,** Chemical and physical methods of enhancing transdermal drug delivery, *Pharm. Tech.,* 12, 130, 1988.
17. **Smith, K. L. and Lonsdale, H. K.,** Membrane systems: theoretical aspects, *Methods Enzymol.,* 112, 495, 1985.
18. **Franz, T. J.,** The finite dose technique as a valid in vitro model for the study of percutaneous absorption in man, *Curr. Probl. Dermatol.,* 7, 58, 1978.
19. **Woodford, R. and Barry, B. W.,** Penetration enhancers and the percutaneous absorption of drugs: an update, *J. Toxicol. Cutan. Ocul. Toxicol.,* 5, 167, 1986.
20. **Knutson, K., Potts, R. O., Guzek, D. B., Golden, G. M., McKie, J. E., Lambert, W. J., and Higuchi, W. I.,** Macro- and molecular physical-chemical considerations in understanding drug transport in the stratum corneum, *J. Contr. Rel.,* 2, 67, 1985.
21. **Idson, B. and Behl, C. R.,** Drug structure vs. penetration, in *Transdermal Delivery of Drugs,* Vol. 3, Kydonieus, A. F. and Berner, B., Eds., CRC Press, Boca Raton, FL, 1987, 85.
22. **Zatz, J. L. and Dalvi, U. G.,** Evaluation of solvent-skin interaction in percutaneous absorption, *J. Soc. Cosmet. Chem.,* 34, 327, 1983.
23. **Chien, Y. W., Keshary, P. R., Huang, Y. C., and Sarpotdar, P. P.,** Comparative controlled skin penetration of nitroglycerin from marketed transdermal delivery systems, *J. Pharm. Sci.,* 72, 968, 1983.

24. *The Merck Index,* Tenth Edition, Windholz, M., Ed., Merck and Co., Rahway, NJ, 1983.
25. *United States Pharmacopeia,* 21st revision, U.S. Pharmacopeial Convention, Inc., Rockville, MD, 1984.
26. *Martindale, The Extra Pharmacopeia,* Reynolds, J. E. F., Ed., The Pharmaceutical Press, London, 1982.
27. **Hansch, C. and Leo, A.,** *Substituent Constants for Correlation Analysis in Chemistry and Biology,* John Wiley and Sons, New York, 1979, 159.
28. **Baker, R. W. and Lonsdale, H. K.,** Controlled release: mechanisms and rates, in *Controlled Release Technologies: Methods, Theory, and Applications,* Tanquary, A. C. and Lacey, R. E., Eds., Plenum, New York, 1974, 15.

Chapter 3

INSTRUMENTATION AND METHODOLOGY FOR *IN VITRO* SKIN DIFFUSION CELLS IN METHODOLOGY FOR SKIN ABSORPTION

Stephen W. Frantz

TABLE OF CONTENTS

I. INTRODUCTION

Penetration of the skin by exogenous materials is an extremely important exposure route for toxicological testing paradigms. The concept of measuring skin penetration by drugs or chemicals by using small pieces of skin removed from animals or humans is not a new one. As early as 1944, Burch and Winsor[1] reported the use of excised skin to conduct *in vitro* and *in vivo* comparisons of transepidermal water loss. However, it has only been within the last decade that the study of *in vitro* skin penetration has benefited from the availability of more sophisticated, automated and commercially available diffusion chambers.

While many variations exist, there are two basic designs which have evolved into practical use: (1) the static, or nonflowing cell and (2) flow-through cells. The basic design of these chambers includes some type of clamping system to position the skin and to provide a liquid-tight seal over a receiver chamber. The receiver fluid in the chamber below the skin may vary in composition but is typically an aqueous nutrient media containing the critical amino acids, salts, and buffers necessary to maintain the excised skin over the duration of the measurement; the reader should refer to the other chapters of this book to provide a more detailed discussion of these receiver fluids. Most diffusion cell designs attempt to mimic the physiologic and anatomic conditions of skin *in situ*. The skin can thus be thought of as a plane of tissue situated over a network of capillaries which bathe the underside of this plane. The skin thus represents a "barrier layer" through which drugs and chemicals must pass in order to be taken up systemically *in vivo*. The *in vitro* diffusion chamber attempts to resemble this tissue layer/capillary network arrangement by positioning a skin sample over a medium which takes up the penetrant chemical. This is particularly true for the flow-through variety of diffusion chambers since media continually flows beneath the skin to mimic capillary blood flow. The penetrating chemical can then be assayed in the receptor medium, either as radioactively labeled test material by radiometric methods or as nonlabeled chemical by standard chromatographic techniques. The other basic design variation is the static, or nonflowing, skin chamber. This chamber type can also provide useful measurements and its design considerations will be discussed in the succeeding pages.

This chapter will provide the reader with a full appreciation of the practical considerations involved in the selection and use of a properly designed skin penetration chamber. First, the various chamber types and fabrication materials which are available will be presented. Second, chamber volume and temperature considerations and differences in skin surface exposure conditions will be discussed. Finally, commercial sources for skin penetration chambers will be provided.

II. DIFFUSION CELL DESIGN CONSIDERATIONS

A. STATIC DIFFUSION CELLS

This design for *in vitro* skin penetration cells was the first to be used to measure absorption of a test material through excised skin. Several variations of this design have evolved and been improved upon over the years but each design contains two chambers which can be classified as either: (1) air/fluid (ambient) phase or (2) fluid/fluid phase chambers. The air/fluid (ambient) phase chambers attempt to leave the epidermal surface of the skin preparation completely uncovered and exposed to room atmosphere conditions while maintaining the dermal surface of the excised skin on some type of organ culture medium or normal saline solution. It can be viewed conceptually as an air/fluid phase system with the skin serving as the multilayered membrane interface.

The fluid/fluid phase diffusion cells were also developed for the purpose of measuring the translocation of drug molecules through skin. However, in this case the translocation is measured as drug passing from one fluid phase (e.g., ointment and cream topical therapeutic

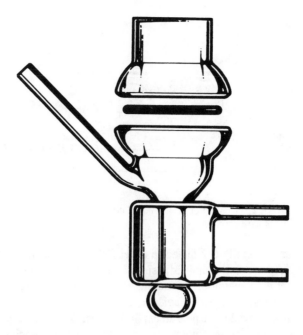

FIGURE 1. Schematic representation of the standard, original diffusion cell developed by Franz.[2] This cell is commercially available in several orifice diameters (5 to 25 mm) and in clear or amberized glass.

applications) through skin preparations into a second fluid phase, which represents the uptake by the capillary beds in the upper layers of the dermis. From a physicochemical point of view, transport between fluid phases should follow Fick's First and Second Laws of Diffusion and the Law of Mass Transport as driving forces for movement from one phase to the next. Thus, as *in vitro* approaches to skin penetration were developed, the fluid/fluid measurements made the most sense for *in vitro* evaluation of drug penetration of the skin, since nonoccluded skin measurements are not particularly relevant for topically applied pharmaceuticals. However, similar principles can be applied with air/fluid systems in which the test penetrant is applied in an ointment (e.g., petrolatum).

Further discussion of ambient vs. fluid phase exposures of excised skin will be presented in a latter segment of this chapter. The differences and similarities of these two basic design types for static diffusion cells will now be discussed.

1. Air/Fluid Phase Static Diffusion Cells

One of the most widely-recognized static design *in vitro* diffusion cells is the Franz diffusion cell, developed by Dr. Thomas J. Franz.[2] This cell has a static receiver solution reservoir with a side-arm sampling port design (Figure 1). It was originally developed for finite dose application[3] of compounds to the epidermis. As seen in Figure 1, the skin is positioned between two halves of a glass chamber. The upper half of the chamber is an open cap, allowing ambient exposure and access to the epidermis for dosing or other purposes. A Teflon® O-ring is used to provide a pressure seal at the periphery of the skin preparation and the two halves are held together with a clamp. The thermal jacket is positioned around the receiver chamber and is heated with an external circulating/bath. A Teflon®-coated stirring bar is placed in the receiver chamber and a physiologic saline solution (or other appropriate medium) is used as a receptor solution to fill this chamber. During the course of an experiment, small volumes of the receptor solution are drawn from the chamber for analysis and are replaced to keep the volume of solution constant during the experiment.

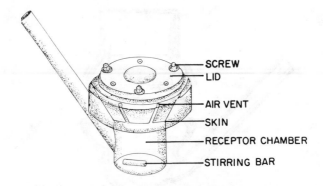

FIGURE 2.   Drawing of the assembled diffusion cell used for Lofts-son's *in vitro* experiments.[5]

Measurements are generally conducted in at least duplicate or triplicate determinations per experimental condition being examined, using sets of calibrated, matched diffusion cells.[3] The reproducibility of the penetration measurement or the variation with experimental conditions can therefore be systematically evaluated. However, when Franz cells have been modified to incorporate large skin surface areas, these larger cells have experienced problems with temperature gradients and incomplete mixing.[4] Such inconsistencies need to be anticipated and corrected by the investigator when design modifications are tested. To evaluate the reproducibility of penetration measurements, multiple static cells can be run in uniform console arrangements (e.g., 8 to 9 cell holders) containing stirring motors for mixing and heated water manifolds for temperature control. The Franz cell has also been modified to include a version which allows for measurement of evaporative loss from the skin surface by volatile chemicals and one which has a flow-through receptor fluid design. These versions will be presented and discussed in subsequent segments of this chapter.

2. Air/Fluid Phase Chambers Related to the Franz Cell Design

Several variations of the static reservoir, side-arm Franz cell exist. Some cells are of much simpler design and essentially represent a skin disk placed over a receiver chamber. One such design (Figure 2) was presented in a paper by Loftsson.[5] The cell was of Plexiglas® construction, with a large receptor chamber (45 ml capacity), a side-arm sampling port design, a lid with a rubber gasket, and a chamber opening which provided an exposed area of epidermis of approximately 8 cm². A Teflon®-coated magnetic stir bar was also used in this chamber. Air vents were incorporated to assure contact of the skin with the receptor phase. The temperature of the receptor fluid was maintained at 31°C by an incubator. A similar cell design was used in work conducted by Nacht et al.[6]

Another similar design for the static, side-arm cell type was reported by Dugard et al.[7] (Figure 3A). This chamber was used for horizontal membrane applications in which a smaller, 1.8 cm² area of the epidermis is available for chemical application. A second, vertical membrane version of this chamber (Figure 3B) will be discussed in the fluid/fluid applications portion of this chapter.

Kao et al.[8] reported the use of an organ culture dish containing a center well (Falcon Labware, Oxnard, CA) for incubation of excised skin samples. Even though this technique was used to incubate skin preparations for toxicity tests, the technique could also be used as a short term screening method for skin penetration. The center well has a capacity for approximately 8 ml of media, and the penetration of skin samples supported on filter paper disks can be quantified in this well volume for short periods of time (e.g., minutes to hours). The sample/filter disk combinations are placed epidermal side up onto the center well of these dishes which have previously been filled with the medium of choice, taking care to

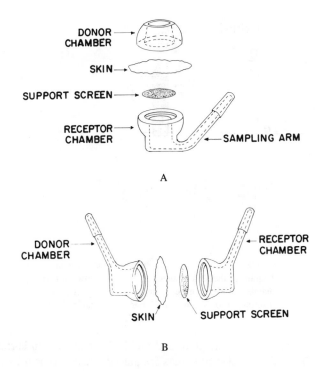

DONOR CHAMBER

SKIN

SUPPORT SCREEN

RECEPTOR CHAMBER

SAMPLING ARM

A

DONOR CHAMBER

RECEPTOR CHAMBER

SKIN

SUPPORT SCREEN

B

FIGURE 3. Glass diffusion cells of two types, (A) air/fluid static design, and (B) fluid/fluid static cells designed by Dugard et al.[7]

avoid air bubbles between the filter paper and the medium. This method may also be used to study the incorporation of radioactive biochemical markers, such as ^{3}H-thymidine incorporation into DNA and ^{14}C-leucine incorporation into protein, as indicators of the viability of excised skin preparations. Simple but effective initial evaluations can be conducted in this manner. Caution should be exercised though, to avoid the overinterpretation of results from this preliminary assay method. It is probable that at least some of the test material may diffuse across the skin surface to the edges of the preparation and some accumulation in the filter support might occur.

3. Fluid/Fluid Phase Static Diffusion Cells

A fluid/fluid phase receptor/donor system treats the skin preparation as a membrane between two chamber halves which are well-stirred, fluid-filled compartments. This design has been referred to as a horizontal, or infinite dose,[3] closed cell type of diffusion chamber. However, it should be noted that it is also possible to apply an infinite dose with an air/ fluid chamber design. In the horizontal type of cell a test material is introduced to the donor compartment. Absorption through the skin is then evaluated by serial sampling of the receiver compartment for changes in test material concentration at several time points following addition to the donor side of the chamber. The data can be expressed as either cumulative absorption vs. time or absorption rate vs. time (Figure 4). The amount of test material which is removed from the donor chamber by skin absorption is typically very small relative to the large amount placed in the donor side. Thus, the dose can be considered infinite relative to the receiver compartment during the short duration of such *in vitro* measurements. Several cells have been developed for these types of measurements.

One example of such a two-reservoir diffusion cell (Figure 5) is the Side-Bi-Side® cell which was reported by Flynn et al.[9] to study skin permeation kinetics. This cell is constructed completely of glass to allow viewing of all aspects of the skin mounting. Each cell half is

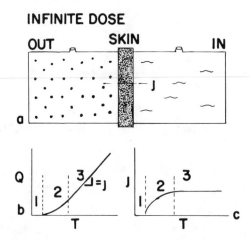

FIGURE 4.   The infinite dose technique (A) Schematic diagram of the skin mounted as a barrier between two chambers. Data can be expressed as either: (B) total absorption (Q) vs. time, or (C) rate of absorption (J) vs. time.

calibrated to 3.4 ml at the sampling stem and yields 2.5 cm^2 of skin surface area for test material transfer. Teflon®-coated stirring bars are placed in each compartment and a thermal jacket heats both sides equally from an external, temperature-controlled water bath. The skin/membrane is protected from tearing or twisting by a special lock nut and clamp assembly.

A similar two-reservoir chamber design (Figure 6) has been reported by Tojo et al.[10] which is also useful for studying permeation kinetics. Like the Side-Bi-Side® cell, it is similar in dimension and can be used for finite or infinite dose applications under isothermal or nonisothermal conditions. It contains two cell-halves which are designed to provide a large skin exposure surface (0.64 to 2.5 cm^2). The two halves are held together with a pinch clamp to provide a leak-tight system. Each compartment can hold up to 3.5 ml of solution, which is calibrated on the sampling stem. Stirring of compartment contents is achieved with Teflon®-coated, star-head magnets which reduce solution drag; these magnets are driven at a constant, synchronous speed of 600 rpm with drive motors located in a compact drive unit. The enclosed half-cell system is designed to minimize loss of any bathing solution volume due to evaporation.

Still another variation of this fluid/fluid cell design has been reported by Dugard et al.,[7] which is an analogous system to the horizontal membrane diffusion cell also reported by these investigators (Figure 3B). This two-reservoir cell utilizes a skin specimen supported on a porous metal screen clamped between a donor and a receptor chamber. The flux, or drug permeation, through the skin/membrane is determined by serial sampling of the receptor compartment. Receptor fluid aliquot volumes are replaced with fresh, blank medium.

B. FLOW-THROUGH DIFFUSION CELL DESIGNS

In contrast to sampling from the static cells, for which serially drawn receptor fluid volumes are replaced manually, the flow-through system can provide automatic replenishment of receptor fluid. Early versions of these flow systems have been described by Ainsworth[11] and Marzulli et al.[12,13] The flow volumes can be regulated by the use of variable flow rate pumps which are adjusted in relation to the fixed size (volume) of the receptor chamber in the diffusion cell. The flow-through diffusion cell also allows the system to be automated in terms of sample collection, offering the advantages of uniform sample collection and unattended operation. Technical problems with the use of these new sytems will be discussed in latter segments of this chapter. Examples of flow-through systems currently in use are now presented.

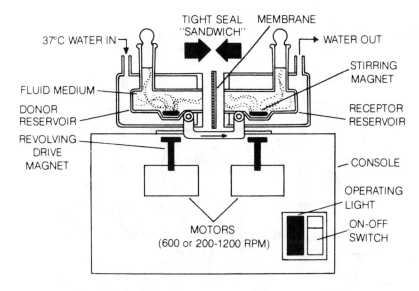

FIGURE 5.   Fluid/fluid static diffusion cell designed by Flynn[9] (commercially available in several sizes).

1. The Flow-Through Franz Diffusion Cell

A flow-through modification of the standard Franz diffusion cell designed by Mathias[14] was constructed entirely of glass and featured a lower port for flow-through applications (Figure 7). Both ports are fitted with glass, luer-lock syringe fittings to allow a conversion to an automatic system. Overall, the cell is much like its nonflow-through predecessor and can be used as either a static or flow-through cell. However, Gummer and co-workers[4] caution that flow-through Franz cells may experience problems with an inability for the receptor volume to be stirred quickly and uniformly. This is particularly true in the "tulip-shaped" bowl design for these cells (see Figure 7).

An alternate flow-through conversion of the Franz cell was reported by Hawkins and Reifenrath.[15] This cell was modified from the standard (static) Franz cell to include a modified sidearm which allowed continuous flow. A tube, fed down the side arm, extends into the body of the cell. The tube is sealed to the side arm and fresh receptor fluid is pumped into the cell via the tube. Receptor fluid containing the penetrant is forced to exit the cell via a "T" in the side arm. Temperature in the receiver well was kept constant with water pumped through the outer jacket of the cell from a 37°C reservoir. This modified cell was developed for use with an evaporation cell to quantify evaporative loss and penetration concurrently. The reader should refer to section III.C. of this chapter for a further discussion of evaporation cells for *in vitro* diffusion cell applications (Figure 13). The skin can be fixed in this penetration cell with an O-ring while the upper portion of the chamber is clamped over the skin preparation. This penetration cell has a 3 ml (nominal) volume for the receiver chamber, and can be conveniently mounted in stainless steel stirring units.

2. The Oak Ridge National Laboratory Skin Permeability Chamber

Holland et al.[16] reported on a multi-sample apparatus to study the *in vitro* penetration of living skin by chemicals. This apparatus, referred to by these investigators as the ORNL skin permeability chamber (Figure 8), was designed to: (1) be easy to use; (2) maintain tissue viability; (3) measure the rate of penetration; and (4) determine the character of materials following penetration of intact skin. This chamber is fabricated from transparent plastic (Plexiglas®) and consists of two major parts, a cover plate and a tissue chamber. The cover plate contains a series of five openings which allow access to the epidermal

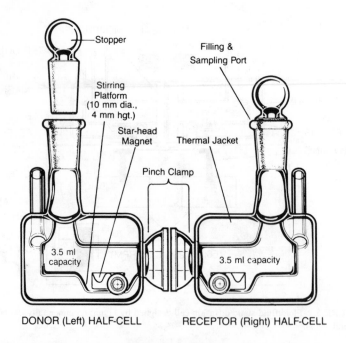

FIGURE 6. Fluid/fluid static diffusion cell reported by Tojo et al.[10] (commercially available in several sizes).

surface of the skin in the assembled chamber. These openings align with the counterbores in the tissue chamber. These openings are fitted with Teflon® bushings which project slightly from the underside of the cover plate. The bushings apply sealing pressure around the perimeter of the 1-in.-diameter skin discs which are supported on a slight depression around the edge of each counterbore; this depression provides a countersurface against which the sealing pressure is applied from the Teflon® bushings.

Receptor fluid temperature is controlled by a circulating hot water bath. Each well is filled with culture medium and the skin discs are positioned in the tissue chamber with the epidermal side up and the dermal side facing down over the culture medium wells. The skin is positioned carefully to exclude air bubbles between it and the volume of culture medium it is placed over. The cover plate is placed over the five skin preparations in a uniform manner to avoid bubble formation and is attached with wing nuts as shown (Figure 8). Holland et al.[16] reported that diffusional lag, or delay in the appearance of test chemical in the chamber effluent due to poor mixing in the medium well, was minimized in this diffusion cell by a cross-flow design of the inlet and outlet ports in the lower chamber. In combination with the small receptor volume (approximately 1 ml), this flow pattern should minimize this delay. As with all flow-through type diffusion cells, the number of turnover times per unit time for the receptor volume is dependent on the size of the receiver well (ml), which is fixed, and the flow rate (ml/h) in and out of the well, which can be varied. The influence of flow rate on *in vitro* skin penetration measurements will be discussed in Section IV of this chapter.

Receptor fluid is pumped from a reservoir by a multichannel cassette-type peristatic pump to inlet ports of the chamber. Effluent receptor fluid is directed to a fraction collector for the automated collection of interval samples and from this data a time-course characterization of skin penetration is obtained.

3. The Flow-Thru® Diffusion Cell

A single-chambered, flow-through diffusion cell was reported by Bronaugh and Stewart.[17] This cell is fabricated from Teflon® and Kel-f®, a semitransparent material used for

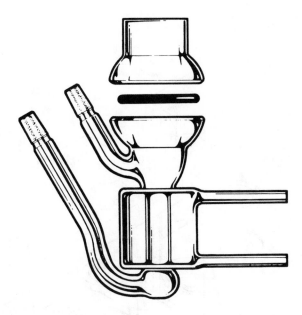

FIGURE 7. Schematic representation of a flow-through variation of the standard Franz diffusion cell designed by Mathias.[14] This cell is commercially available in several orifice diameters (5 to 25 mm) and in clear or amberized glass.

the side arms and the cap for this cell (Figure 9A). Teflon® was selected for construction of the cell body because it is inert and unbreakable. The cell has a glass window on its bottom surface for viewing air bubbles below the skin sample, which can be removed by a simple tilt of the cell. The skin is held in place by a two-piece upper nut (cap). This cap is comprised of an outer, threaded nut made of Kel-f® which screws into the cell body, and an inner, free-moving Teflon® bushing to hold the skin preparation in place without tearing or twisting. After assembly, the cap is open to the air so that the skin can be dosed. The entire cell is very compact, with an exposed surface area of 0.32 cm² and a receptor volume of 0.13 ml. The cell's outlet sidearm has a smaller inner diameter than the inlet sidearm, creating a slight back pressure as the receptor fluid exits the chamber. This back pressure is intended to promote greater contact between skin and the receptor fluid and facilitate uptake of the diffusing chemical. However, the investigator should exercise caution so that excessive back pressure is avoided. Overall, the cell's design was intended to provide continuous replacement of receptor fluid, good mixing of receptor contents, and rapid removal of the penetrating test chemical.

During chamber operation, the individual cells can be inserted into a 7-position, anodized aluminum heater block which maintains the temperature in the cell, typically at 32°C (Figure 9B). The related equipment for the system is very similar to the one described earlier for the ORNL skin permeability chamber.

4. The LGA Diffusion Cell Block System 4 Flow-Through Chamber

The author and co-workers have developed a custom-fabricated, flow-through design skin chamber[18] (Laboratory Glass Apparatus, Berkeley, CA, Figure 10). Its basic design was taken from the ORNL skin chamber[16] (Figure 8), but it is larger overall with several features introduced to improve upon the performance of the previous apparatus.

Most of the chamber is fabricated from transparent acrylic (similar to its ORNL predecessor) to be of light, yet sturdy construction. However, the receiver well walls of the chamber are lined with Teflon® and each well bottom is lined with glass (Figure 11). With

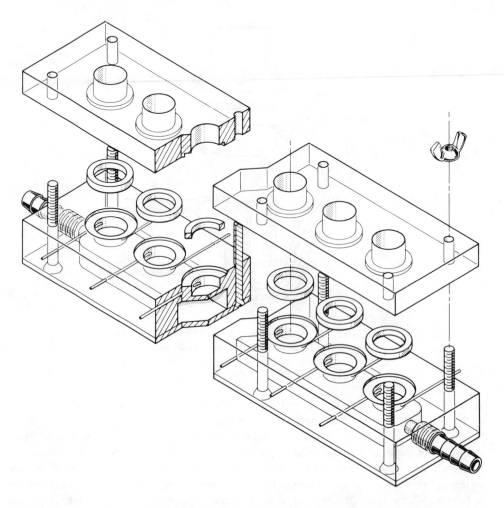

FIGURE 8.   Schematic view of the ORNL skin permeability chamber.[16] This flow-through design chamber was produced in several versions, containing from 5 to 10 individual wells for maintaining skin preparations under uniform experimental conditions. A Teflon® cap (not shown) was also available in this chamber for use with volatile chemical applications.

the exception of the stainless steel inlet and outlet (culture media) ports, the test chemical and the effluent medium contact only Teflon® and glass. The openings in the acrylic upper plate are also lined with Teflon® in a specially-machined O-ring bushing. The modified bushing provides a continuous Teflon® surface from O-ring (at the skin surface) to the top of the upper plate which ends in a tapered opening, as seen in cross-sectional view (Figure 11). This opening will accept an evaporation cell with a ground-glass joint (Laboratory Glass Apparatus), which is similar to the evaporation cell described by Hawkins and Reifenrath[21] (Figure 13), or a glass stopper, thus giving the investigator several experimental options.

The contents of each receiver well (approximately 1.7 ml volume) are stirred by Teflon®-coated magnetic stirring bars (Figure 11 inset), which are driven at a constant, synchronous speed (600 rpm) by magnetic heads on compact drive motors located beneath the lower surface of each receiver well. The magnetic heads are in close proximity with the stirring bars by projecting into a recess in the manifold which runs beneath and around each well (Figure 11, lower drawing). Temperature in the block is maintained by this manifold, which runs the length of the chamber and is fed continuously from a circulating water bath. As with the ORNL chamber, at the start of an experiment the upper plate is carefully placed

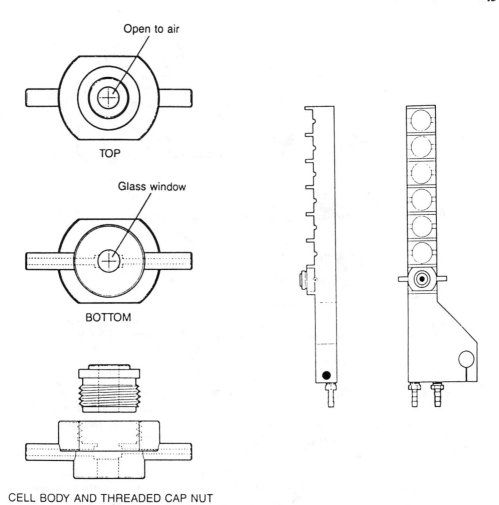

FIGURE 9. Schematic drawing of the (A) flow-through diffusion cell and (B) holding block reported by Bronaugh and Stewart.[17] Up to seven cells at a time can be used with the holding block which is maintained at constant temperature with heated water from a circulating water bath.

onto six skin preparations simultaneously and secured with wing nuts to provide leak-tight seals around the skin discs. Care must also be taken to avoid the introduction of air into the receiver wells as the skin preparations are positioned and the upper plate is secured. Effluent media is collected over preselected intervals in a fraction collector for later analysis.

C. DIFFUSION CELLS FOR THE MEASUREMENT OF VOLATILE CHEMICALS

A number of modified diffusion cells have been introduced to measure the penetration of volatile chemicals. However, only a few cells are available which allow the simultaneous quantification of evaporation and penetration processes. The simplest design modifications have involved the addition of a special cap for the diffusion cell, which would otherwise be left open to the air. Such caps can be of a very simple nature, like the ground joint stoppered cap for the Franz diffusion cell (Figure 12A), the Teflon® stopper described for use with the ORNL flow-through chamber (in Figure 8) or the ground glass stoppers for the LGA flow-through cell (Figure 11 inset). Several different special purpose caps exist for the standard Franz diffusion cell. These include a volatile trap cap (Figure 12B), which

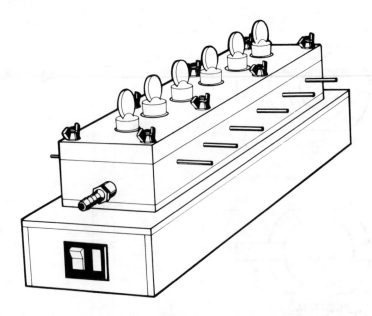

FIGURE 10.   The Diffusion Cell Block System 4® flow-through chamber designed by Frantz.[18] Six skin preparations can be simultaneously evaluated in this chamber under uniform experimental conditions.

contains a threaded nylon nut to hold a charcoal tube for trapping of volatile organics, or an airflow version of the volatile trap cap that also uses a charcoal trap for collection of volatiles; other air flow versions of these caps exist as well for the introduction of simple air flow into a closed system (Figure 12C).

The simultaneous measurement of evaporation and skin penetration has been extensively studied by Spencer et al.[19] and more recently, has been described by Hawkins and Reifenrath.[20-22] The modified penetration cell approach that was used in these papers was discussed previously (Section II.B.1) and was adapted from a previous design for an all-glass evaporation/penetration cell used by Spencer et al.[19] The measurement of the concomitant, evaporative loss of test chemical was accomplished using an evaporation manifold with a replaceable vapor trap (for interval sampling) and an air control system. This manifold, which was described in an earlier paper by the senior author,[21] was modified by the addition of a water jacket to improve temperature control (Figure 13A). The opening on the replaceable vapor trap is positioned approximately 9.5 mm from the surface of the skin and contains Tenax GC®, or another suitable trapping sorbent, to quantify evaporation rates and extent. The entire penetration-evaporation cell (LGA Skin Permeation System LG-1083-C, Laboratory Glass Apparatus, Berkeley, CA) was recently reported by Hawkins and Reifenrath[22] and is shown schematically in Figure 13B. Further discussion of evaporation cell measurements can be found in the chapter by Hawkins entitled *Methodology for the Execution of In Vitro Skin Penetration Determinations.*

Control of air temperature and humidity is critical for these types of measurements.[22] A fluctuating ambient humidity can be removed as an experimental variable by passing it through a desiccating tower. This air is warmed to experimental temperature with a heat exchanger. The temperature of the evaporation manifold's water jacket is kept isothermal with the heat exchanger. The flow of air is maintained by exhausting the conditioned air through the manifold with a vacuum pump. This flow is controlled by a micrometer valve and monitored with ball-type flow meters at both the intake and exhaust sides of the evaporation manifold. This evaporation manifold, in modified form, can also be used with the LGA chamber, as previously described (Section II.B.4).

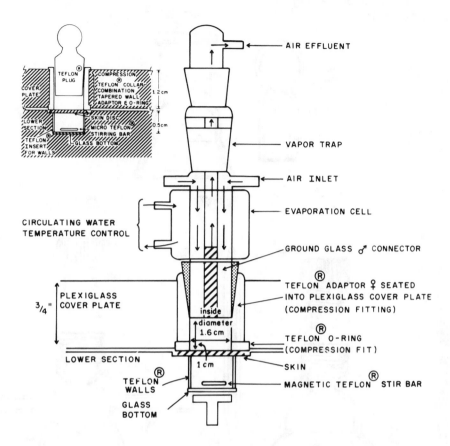

FIGURE 11. Evaporation cell adaptor for use with the Diffusion Cell Block System 4® can be used to measure evaporation vs. penetration of skin (this evaporation cell is a modification of the cell reported by Hawkins and Reifenrath[20]-Figure 13). The block "T" in the lower drawing represents the recessed position of the magnetic stirring motor/magnet assembly. The openings for each cell can alternatively be closed with a glass stopper (inset).

A variation of the Teflon® Flow-Through diffusion cell of Bronaugh and Stewart[17] also exists for volatile compound determination in a sealed system. This modification (Figure 14) is basically an extended version of the threaded Kel-f® cap described previously for the first cell (Figure 9) which includes a Teflon®-faced septum and septum cap to prevent evaporative loss. The test material is delivered through the enclosed septum system (after the skin has been positioned and the upper cap is secured) to prevent loss of the chemical during delivery. In addition, the inlet and outlet sidearms are offset in this version of the cell (cf. top view, Figure 14) to produce a vortex in the receptor well at high flow rates for maximum mixing of the receptor solution. There is also a conversion available from the supplier (Crown Glass, Neptune, NJ) for sidearms which are tapped into the threaded neck if air flow over the skin surface is desired.

III. MATERIALS SELECTION FOR DIFFUSION CELL FABRICATION

The choice of materials for the fabrication of a new diffusion cell, or even for the improvement of an existing cell design, is a basic and very important factor in the successful *in vitro* investigation of chemical penetration of the skin. These cells have been fabricated from several kinds of materials, including stainless steel, glass, Plexiglas®, and Teflon®. The pros and cons of using each of these materials will now be examined.

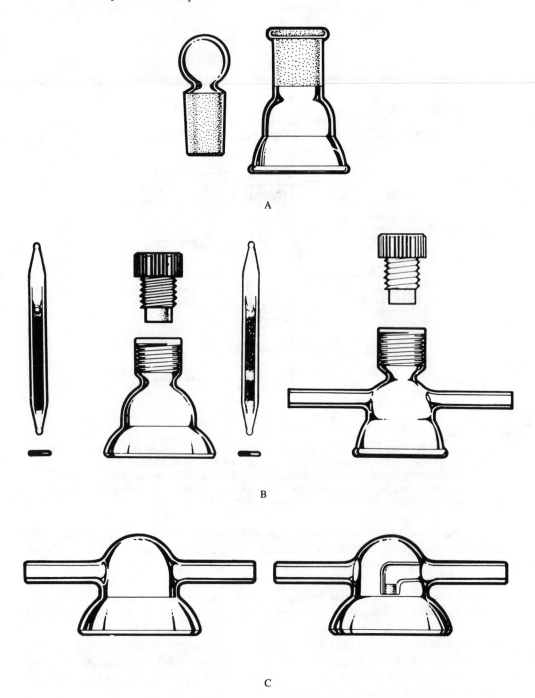

FIGURE 12.   Several optional caps available for use with Franz-type static or flow-through diffusion cells: A) ground joint stoppered cap; B) volatile trap cap (upper) or air flow cap with filter trap (lower); and C) air flow cap (upper) or filtered air flow cap (lower).

49

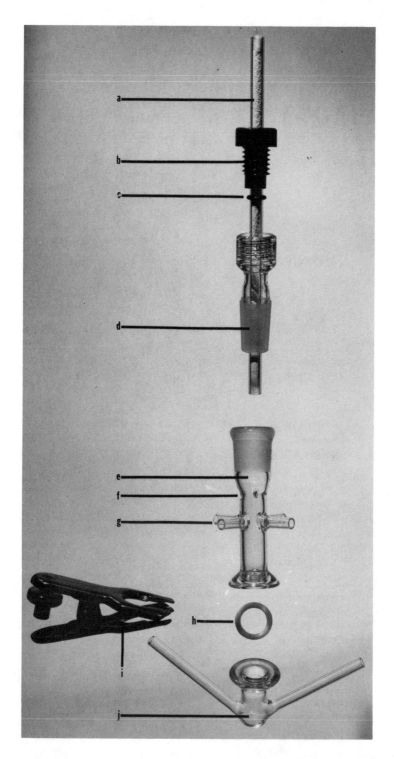

FIGURE 13. *In vitro* skin penetration - evaporation cells: A) Original design of Spencer et al.[19]; and B) design reported by Hawkins and Reifenrath[20] which enables the investigator to simultaneously measure competing rates of loss for test materials from the surface of the skin.

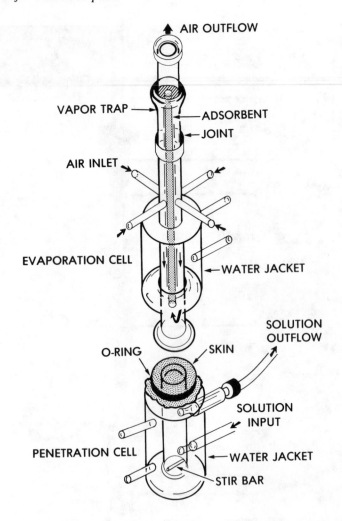

↑ AIR OUTFLOW

VAPOR TRAP — — ADSORBENT

— JOINT

AIR INLET

EVAPORATION CELL — WATER JACKET

SOLUTION
OUTFLOW

O-RING — SKIN

SOLUTION
INPUT

PENETRATION CELL — WATER JACKET

— STIR BAR

FIGURE 13B.

A. STAINLESS STEEL CELLS

This material was used to construct some of the first cells used to measure skin penetration.[12,13] Since it is a heavier material than glass or plastic, it is much more difficult, and can therefore be more expensive, to machine to specifications than are the lighter materials. Diffusion cells made of stainless steel are sturdy and unbreakable, unlike their glass counterparts. However, the opacity of stainless steel makes the observation and removal of unwanted air bubbles in the cells a difficult and often undetected problem. The use of glass windows in the cell can help minimize this technical problem, but the fabrication of more recent chambers has made use of the lighter, transparent materials.

B. GLASS DIFFUSION CELLS

It is clear that glass has been the most commonly used fabricating material for diffusion cells. Glass provides clear views of all aspects of the diffusion cell system and is available in either the standard clear glass or an amberized version, for work with light-sensitive chemicals. It is relatively sturdy, depending on applications, and allows the formation of seamless joints within the cell which articulate with connecting joints in order to provide a greater degree of control over experimental conditions. However, a very real shortcoming in using glass is breakage, which can render a relatively expensive glass apparatus useless

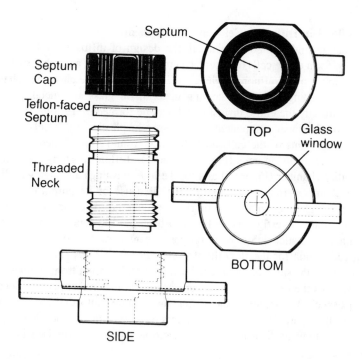

FIGURE 14. The volatile compounds version of the original flow-through diffusion cell reported by Bronaugh and Stewart[17] allows the collection of receptor fluid in a sealed system.

until repairs can be made. Also, glass chambers are often not useful for measuring the skin penetration of silanes. This is due to the adherence of these chemicals to glass although some investigators have used this adherence by silanes to glass to their advantage by pretreating the glass walls of their cells with a "siliclading" agent to prevent adherence by the penetrant. Despite these disadvantages, glass diffusion cells remain the predominant type in present use.

C. PLASTICS IN DIFFUSION CELL FABRICATION

With the advent of hard, easily formed and machined plastics, several new options became available to skin penetration investigators. Plastic afforded the opportunity to make diffusion cells (or chambers containing several cells) easy to use, lightweight, relatively unbreakable, and transparent at critical sites. The principal plastics in current use are Plexiglas® and Teflon®.

1. Use of Plexiglas® in Diffusion Cells and Components

Plexiglas® was introduced as a thermoplastic material which is light, sturdy, permanently transparent and resistant to degradation due to environmental factors. It made an ideal material to form the base structure of a diffusion cell because it is easily machined and compatible with other materials often used in these cells (e.g., metal fixtures, Teflon®, etc.). Because of its resistance to breakage, it is often used to make holding devices for all-glass cells to reduce accidental damage. However, because Plexiglas® is a polymeric material, the major criticism of its use in diffusion cells and multiple-cell chambers is its potential to absorb some organic chemicals, such as solvents, during penetration experiments. This absorption may be followed by leaching into receptor fluids of these absorbed materials from the Plexiglas® in subsequent experiments.[17] This absorption may lead immediately to artifactual losses of the penetrating chemical during the experiment, possibly resulting in lower assessments of penetration than has actually occurred.

2. Use of Teflon® in Diffusion Cells and Components

Teflon® has recently found its place in the design of diffusion cells as a hard, inert, sturdy plastic to form entire cells or to replace ground glass joints and other surfaces. Teflon® is the trademark for polytetrafluoroethylene and is a waxy, opaque plastic which is principally used to prevent sticking, both as a coating material for glass and metal and in industrial applications. Virtually the entire diffusion cell can be formed of Teflon® (Flow-Thru® cell, Figure 9), with additional components of transparent plastic (e.g., Kel-f®) and glass. In other cases, components of the cells are fabricated from Teflon®, such as the Teflon® O-rings for the Franz diffusion cells (Figures 1 and 7), Teflon® bushings used in the ORNL skin permeability chamber (Figure 8) and the Teflon® bushings and side wall linings of the flow-through chamber from LGA (Figure 10). Because of its purported properties, this plastic has been highly regarded as a fabrication material to replace Plexiglas® and glass components by many investigators. It has also been used to replace ground glass surfaces to form fit-tight, non-stick joints between component parts for such modifications as the evaporation cell option for the flow-through chamber from LGA (Figure 11). The opacity of Teflon® presents the same problem cited earlier for fabrication from stainless steel, unless the design includes the use of glass or other transparent materials at critical locations. Such combination usages, for instance, have been incorporated into the design of the Teflon® Flow-Thru® Cell (Figure 9). Some investigators have encountered problems with chemical absorption by Teflon® in diffusion cells, so caution should be exercised in its application for diffusion cells.

D. MATERIAL COMBINATIONS IN DIFFUSION CELL FABRICATION

It should be obvious to the reader at this point that combinations of materials may offer the most flexibility in the design of new diffusion cells or the improvement of existing ones. As lighter, transparent, inert, and sturdier plastic materials continue to evolve, the successful replacement of metal and breakable glass components can occur. However, it is the combination of all or most of these fabrication materials which may lead to more reasonable diffusion cell designs. As examples of this, the ORNL skin permeability chamber (Figure 8), the Flow-Thru® Diffusion Cell (Figure 9) and the flow-through chamber from LGA (Figure 10) all use some combination of metal, glass, plastic and Teflon® in their complete systems.

IV. VOLUME AND TEMPERATURE CONSIDERATIONS

In designing the physical layout of a diffusion cell, several factors should be considered to ensure that the appropriate level of control for experimental variables will be available to the investigator. The most important factors include volume and temperature.

A. GENERAL VOLUME CONSIDERATIONS

Volume, as referred to in regard to diffusion cell specifications, is generally concerned with two compartments in the cell: (1) the donor (or application) compartment and (2) the receiver compartment. For fluid/fluid phase diffusion cells, or so-called horizontal cells, the volume of the donor and receiver compartments is made nearly identical, within reasonable accuracy for glassware tolerances, to facilitate the measurement of penetrant concentrations. On the other hand, the vertical design diffusion cells such as the Franz-type and any of the flow-through cell designs have donor compartments which are not necessarily symmetrical in volume with their counterpart receiver compartments. The only requirement usually imposed on the donor chamber volume is that it be large enough to contain a small volume (typically 1 ml or less) of the applied dosing solution or suspension. For finite dose applications,[3] where the test chemical is distributed on the skin as a thin film in a volatile vehicle which quickly evaporates, the volume of this compartment becomes virtually meaningless.

B. MIXING CONSIDERATIONS IN THE RECEIVER FLUID COMPARTMENT

In order to appropriately quantify skin absorption under standardized, *in vitro* conditions, the molecules of test chemical which penetrate must be taken up by a receptor phase which represents a well-stirred compartment. Fluid/fluid phase (horizontal) diffusion cells require measurements from two separate but experimentally-identical compartments (e.g., fluid type, volume, constant stirring, etc.) to assess penetration. For these cells, mixing is critical and must be maintained throughout the experiment.

Similar considerations are important to the upright diffusion cells. With the Franz cell, and related static, or nonflowing, cell types, complete mixing is crucial to the measurement of penetration rates and the overall characterization of the time course for skin penetration. Many of these cells have relatively large receptor volumes (3 to 7 ml or greater) and rely on a magnetic stirring bar which is impelled by a standard 600 rpm timing motor to affect a homogeneous mixing of the penetrating molecules. In the absence of this homogeneous mixing, removal of the receptor fluid proximal to a sampling port could be misrepresentative of the actual amount penetrating at the time of sample withdrawal. Furthermore, the fluid volume which is removed is then replaced with fresh bathing fluid; this addition would result in lowered proximal solute concentrations upon frequent subsequent sampling in the absence of stirring.

The volume of the receptor well varies for the different available flow-through cells but the concentration of penetrant is dependent on flow rate. The flow rate is adjusted by the user to maintain sink conditions. In general, it should reflect the number of turnover volumes per unit time required to sufficiently remove penetrating molecules. Although blood flow *in vivo* is qualitatively different from flow rate *in vitro* due to the absence of capillary perfusion *in vitro*, flow rate should reasonably approximate blood flow *in situ*. For instance, the flow of blood in human skin is about 3.0 ml/cm^2/h at approximately 25°C;[23] blood flow in the skin of mice and rats is about 0.5 to 1.0 ml/cm^2/h by comparison.[24] Sufficient removal is defined here as the absence of an increase in penetrant chemical concentration in the receptor fluid which affects the diffusion gradient across the skin. This is typically avoided in fluid/fluid cells by providing a receptor volume which is relatively large and which does not substantially increase in penetrant concentration over the short duration of a typical experiment.

For a flow-through system, the flow rate relative to the receptor well volume is critically related. Thus the number of times a receptor well has its contents (total volume) completely replaced in an hour is directly dependent on the flow rate (volume/h) and can greatly influence the apparent time-course from effluent measurement if it is insufficient. However, it should be carefully noted that flow rate alone may not ensure complete mixing of penetrant in the receptor fluid. Depending on the cell type used and the chemical being measured, flow rate adjustments alone may not demonstrate measurable differences. Holland et al.[16] presented evidence for the influence of flow rates varying from 2 to 7.5 ml/h on *in vitro* benzo[a]pyrene recovery after 16 h in the ORNL chamber. Their results showed no statistically significant differences in the total amount of recovered radioactivity with variation in flow rate and these investigators concluded that no particular advantage was gained with flow rates exceeding a minimum of 2.0 ml/h. Subsequent data by Frantz et al.[25] have shown that measurements in this ORNL chamber did not achieve complete mixing in the absence of receptor well stirring. This condition was worsened by lowering the flow rate (1 ml/h) from the standard rate (2.5 ml/h) being used without stirring. Further work by these investigators[25] showed that the addition of magnetic stirring bars and motorized stirring magnets to the system improved the time course for exiting of test material from the receptor to that which can be predicted by standard first-order loss equations for a well-mixed system.[26]

C. DILUTION OF RECEIVER COMPARTMENT CONTENTS: ANALYTICAL CONSIDERATIONS

In order to demonstrate reliably in an *in vitro* study that a test chemical or its metabolite(s) penetrates skin as a particular chemical form, the dilution of chemical in the receptor fluid must be considered. The size (or volume) of the receptor compartment in a static diffusion cell becomes very important. If the volume in this receptor compartment is very large, the penetrant molecules will be quickly mixed into the large solution volume and thus, may become too diluted to be detected by the analytical techniques available; this may be particularly troublesome with slowly penetrating chemicals. Similarly, a small receptor volume and a large flow rate in a flow-through cell can result in a relatively small amount of penetrant chemical being dissolved in a very large volume of receptor fluid channeled to a fraction collector; this dilution problem will be enhanced with long collection intervals. Such a circumstance can make even the usually sensitive radiometric determinations a difficult prospect for the very slowly penetrating test chemicals. Careful selection of a diffusion cell type with the dimensions which are most appropriate to the particular application is required. In the use of a flow-through cell, selection of an appropriately reduced flow rate will generally provide reasonable concentrations to measure. In addition, caution must be exercised in reducing this flow rate when working with skin preparations which require certain levels of oxygen and nutrients to remain viable over the course of the measurement (e.g., *in vitro* skin metabolite identification studies). In these cases, it might be possible that a reduced flow rate could influence the rate of penetration if critical levels for nutrients, oxygen and pH control factors were reduced beyond physiological lower limits. For such applications, the relationship between these limits and flow rate clearly would need to be established for the cell, the skin type used, and the particular chemical under study.

One of the principal advantages to using *in vitro* diffusion cells is that the identity of the penetrating chemical (e.g., applied chemical vs. another chemical moiety) can be more easily investigated than is often the case for *in vivo* measurements. This is because plasma concentrations are often too low to be detected in an *in vivo* determination due to the relatively poor limits of quantitation which may occur and the small plasma sample volumes which may be available. With an *in vitro* skin penetration study, the opportunity exists to measure penetration of the test chemical in the absence of factors which may confound the interpretation of results (e.g., plasma protein binding). The *in vitro* technique may also be useful to study a "first pass" metabolism effect in skin for a selected number of compounds, as a growing number of researchers have recently reported.[5,6,16,27-36]

D. TEMPERATURE CONSIDERATIONS

Temperature control is an important factor in the minimization of variation in experimental conditions. Many of the diffusion cells in use at present have water jacket-type control of temperature so that penetration measurements can be conducted under uniform temperature throughout the experiment. This type of temperature control also offers the investigator the option to vary the temperature uniformly in order to conduct systematic evaluations of the influence of temperature changes on skin penetration for a particular chemical.

1. Skin Surface and Receptor Fluid Temperature

The selection of experimental temperatures for *in vitro* skin penetration measurements has not been uniform among researchers. Some have chosen to conduct their measurements with the receptor fluid maintained at 37°C to be consistent with the body's core temperature.[2,16,20,22,25] Other investigators have selected 31 or 32°C as their receptor fluid temperature,[5,17,21] presumably to coincide with *in vivo* measurements of the surface temperature of skin.[37] Still others conduct their determinations at (or near) room temperature (approximately 25°C).[7] Many investigators have focused on 32°C for the receptor fluid as the standard temperature for *in vitro* percutaneous absorption measurements.[17,18] This is consistent with a recent recommendation from consensus guidelines for percutaneous *in vitro* studies to maintain the surface temperature of skin at 32 ± 1°C.[38] However, variation of the temperature of the skin to study the influence of this variable on skin penetration can be an advantage in using *in vitro* diffusion cells. It may be beneficial in initial investigations with new, untested chemicals to consider evaluating the effect of skin temperature on skin penetration.

2. Temperature Control of the Bathing Solution

Control of receptor fluid temperature (buffer, saline, culture medium, etc.) is usually maintained with a water jacket or other form of thermal convection. Care must be taken by the investigator to keep the surface temperature of the skin at the normal physiologic condition, since temperature elevation from the receiver compartment may lead to the increased hydration of the skin.[39-41] This elevated hydration condition may tend to preferentially favor penetration of hydrophilic chemicals and reduce the penetration of more hydrophobic/lipophilic chemicals.

Temperature increases as receptor fluid passes from the reservoir to the diffusion cell receiver compartment may lead to the undesirable formation of air bubbles within the diffusion cell due to loss of dissolved gas. This bubble formation may possibly contribute to variation in the measurement between replicate skin samples. This is particularly true for the flow-through diffusion cells, especially if high flow rates are utilized. With these cells, the medium is typically pumped from a reservoir maintained at room temperature into a water-jacketed, heated compartment (e.g., 32 or 37°C). This can be remedied by maintaining the reservoir temperature at a value of approximately 4 to 5°C higher than temperature maintained in the diffusion cells; this temperature elevation should not be detrimental to the particular medium in use. Thus, even with some temperature loss as the fluid is pumped from reservoir to diffusion cell, the temperature will still be slightly higher as the fluid enters the diffusion cell. Any dissolved gases in the medium, particularly when the medium is oxygenated to maintain skin viability, will not tend to leave the solution as the temperature slowly drops.

3. Influence of Receptor Fluid Temperature on Penetration Rates

It is known that a 10°C rise in the temperature of skin can produce a 2- to 3-fold increase in permeability by decreasing the activation energy required for diffusion.[23] Only a limited number of reports are published which detail systematic investigations of the relationship between temperature changes and skin penetration. Several of these investigations were concerned with the concomitant effects of temperature and humidity fluctuations. Fritsch and Stoughton[39] reported that a tenfold increase in the penetration of excised human (autopsy) skin was observed for both acetylsalicylic acid and glucocorticosteriods when the environmental temperature was increased from 10 to 37°C. Stoughton[40] showed enhanced penetration of ethyl nicotinate with both warm and cool water immersion of skin. Studies by Blank et al.[42] demonstrated that increases in temperature from 5 to 50°C dramatically increased the permeability constants of the C_4-C_8 straight-chain aliphatic alcohols using excised human (autopsy) skin. Feldmann and Maibach[43] increased the temperature from 10 to 40°C for excised human (autopsy) skin while maintaining relative humidity at 50%; these workers found that the penetration by ^{14}C-acetylsalicylic acid increased about 15-fold with this increase in temperature. In another study, Wang et al.[44] showed that rabbit skin *in vitro* was more rapidly penetrated at 34 ± 1°C by both a water-soluble padan derivative and the fat-soluble permethrin than at 24 ± 1°C. Similar findings have been reported by Hawkins and Reifenrath[20] for effects of temperatrue on both skin penetration and evaporation of test chemicals; these two factors are inversely related (increased penetration and decreased evaporation) as temperature is increased from 20 to 32°C. These limited examples illustrate that the influences of temperature changes on skin penetration may be dramatic and can be readily evaluated using *in vitro* diffusion cell methodology.

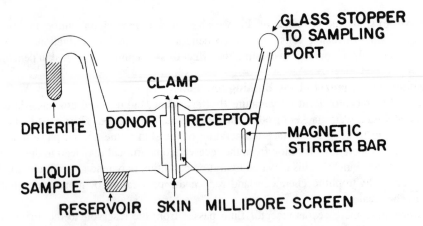

FIGURE 15.   Glass cell for the measurement of vapor diffusion through skin as reported by Harrison et al.[46]

4. Ambient vs. Occluded Exposure of Skin *In Vitro*

Several diffusion cells allow penetration measurements to be conducted under ambient conditions. These include the Franz cell,[2] the static diffusion cells reported by both Loftsson[5] and Dugard et al.,[7] and the ORNL,[16] Flow-Thru®,[17] and LGA[18] flow-through penetration chambers. These cells possess openings which permit the application of test chemicals without occlusion. This feature is used to simulate occupational exposure of chemicals to normal, nonoccluded skin. It is recognized that occlusion may enhance penetration and exaggerate the hazard.

Other investigations may require occlusive or semiocclusive conditions to study drug/chemical penetration, particularly with volatile chemicals or dosing vehicles. The Franz cell includes several options for occlusion (cf. Figure 12) and both the ORNL chamber[16] (Figure 8), Flow-Thru® cell (Figure 9), and the LGA chamber[18] (Figure 11) can be used with stoppers to occlude the skin preparation. The evaporation cells previously described[20-22] allow a variation in the air flow rates in order to study the effect of occlusion vs. ambient conditions.[45] In these evaporation cells, the air flow patterns can be visualized and adjusted with the use of hand-held "smoke guns", which are commercially available. A variation of these applications was reported by Harrison et al.[46] for vapor diffusion through skin. This diffusion cell (Figure 15) resembles the fluid/fluid cell type. The donor compartment is partially open to the room air through a drying trap but contains a well for placement of a volatile liquid. Thus, the penetration of skin by organic vapors can be measured in this manner.

The fluid/fluid phase diffusion cells[7,9,10] represent an occlusive exposure condition since the skin is completely immersed on both sides by fluid. Materials which are intentionally applied to skin (e.g., topical medications, lotions, etc.) represent occlusive layers. A fluid/fluid arrangement was considered to be a reasonable approximation of this condition.

The variety of apparatus available for *in vitro* skin penetration studies allow the investigator to make penetration measurements using a number of different approaches and experimental conditions. It is thus up to individual researchers to make their own judgements as to the number and types of experimental variables which can be introduced and still result in the collection of meaningful, interpretable data from *in vitro* skin penetration studies.

APPENDIX I. SOURCES FOR THE PURCHASE OF DIFFUSION CELLS

Commercially-available diffusion cells can be purchased from three major sources; the cell chamber types and their sources are

Cell Type	Source/Supplier
Franz cell (and related variations, options, etc.)	Crown Glass, Somerville, NJ
	Vanguard International, Neptune, NJ
Side-Bi-Side® Cell	Crown Glass, Somerville, NJ
Two-Reservoir Cell	Crown Glass, Somerville, NJ
ORNL Skin Permeability Chamber	Oak Ridge National Laboratory
Flow-Thru® Diffusion Cell	Crown Glass, Somerville, NJ
LGA Diffusion Cell Block System 4 Evaporation/Penetration Cell	Laboratory Glass Apparatus, Berkeley, CA
	Laboratory Glass Apparatus, Berkeley, CA

The ORNL chamber was produced at Oak Ridge National Laboratory and may be available for purchase upon written request. The other cells listed can be easily obtained from the supplier indicated and their complete addresses are given in the section below.

In addition to the commercially-available diffusion cells, several firms will fabricate the design and specifications desired for custom production of diffusion cells. The following list is not all-inclusive but represents the major sources known for their custom fabricated work:

- Crown Glass Co., Inc.
  990 Evergreen
  Somerville, NJ 08876
- Laboratory Glass Apparatus, Inc.
  1200 Fourth Street
  Berkeley, CA 94710
- Vanguard International, Inc.★
  1111-A Green Grove Road
  Neptune, NJ 07753

★ Distributor for Crown Glass, Inc.

REFERENCES

1. **Burch, G. E. and Winsor, T.,** Rate of insensible perspiration (diffusion of water) locally through living and through dead human skin, *Arch. Intern. Med.,* 74, 437, 1944.
2. **Franz, T. J.,** Percutaneous absorption: on the relevance of *in vitro* data, *J. Invest. Dermatol.,* 64, 190, 1975.
3. **Franz, T. J.,** The finite dose technique as a valid *in vitro* model for the study of percutaneous absorption in man, in *Current Problems in Dermatology,* Vol. 7, Simon, G. S., Paster, Z., Klingberg, M. A., and Kaye, M., Eds., S. Karger, Basel, 1978, 58.
4. **Gummer, C. L., Hinz, R. S., and Maibach, H. I.,** The skin penetration cell: a design update, *Int. J. Pharmaceut.,* 40, 101, 1987.
5. **Loftsson, T.,** Experimental and theoretical model for studying simultaneous transport and metabolism of drugs in excised skin, *Arch. Pharm. Chem., Sci. Ed.,* 10, 17, 1982.
6. **Nacht, S., Yeung, D., Beasly, J. N., Anjo, M. D., and Maibach, H. I.,** Benzoyl peroxide: percutaneous penetration and metabolic disposition, *J. Am. Acad. Dermatol.,* 4, 31, 1981.

7. **Dugard, P. H., Walker, M., Maudsley, S. J., and Scott, R. C.,** Absorption of some glycol ethers through human skin *in vitro, Environ. Health Persp.,* 57, 193, 1984.
8. **Kao, J., Hall, J., and Holland, J. M.,** Quantitation of cutaneous toxicity: an *in vitro* approach using skin organ culture, *Toxicol. Appl. Pharmacol.,* 68, 206, 1983.
9. **Flynn, G. L., Smith, W. M., and Hagen, T. A.,** *In vitro* transport, in *Transdermal Delivery of Drugs,* Vol. 1, Kydonieus, A. F. and Berner, B., Eds., CRC Press, Boca Raton, FL, 1987, 46.
10. **Tojo, K. and Chien, Y. W.,** Drug permeation across the skin: effect of penetrant hydrophilicity, *J. Pharm. Sci.,* 76, 123, 1987.
11. **Ainsworth, M.,** Methods for measuring percutaneous absorption, *J. Soc. Cosmet. Chemists,* 11, 69, 1960.
12. **Marzulli, F. N.,** Barriers to skin penetration, *J. Invest. Dermatol.,* 39, 387, 1962.
13. **Marzulli, F. N., Callahan, J. F., and Brown, D. W. C.,** Chemical structure and skin penetrating capacity of a short series of organic phosphates and phosphoric acid, *J. Invest. Dermatol.,* 44, 339, 1965.
14. **Mathias, C. G. T.,** Percutaneous absorption: an automated technique for *in vitro* measurements using continuous perfusion diffusion cells, *Clin. Res.,* 31, 586A, 1983.
15. **Hawkins, G. S. and Reifenrath, W. G.,** An *in vitro* model for determining the fate of chemicals applied to the skin, in Proceedings of the 1984 Army Science Conference, West Point, NY, June 1984, 329.
16. **Holland, J. M., Kao, J. Y., and Whitaker, M. J.,** A multi-sample apparatus for kinetic evaluation of skin penetration *in vitro:* the influence of viability and metabolic status of the skin, *Toxicol. Appl. Pharmacol.,* 72, 272, 1984.
17. **Bronaugh, R. L. and Stewart, R. F.,** Methods for *in vitro* percutaneous absorption studies. IV. The flow-through diffusion cell, *J. Pharm. Sci.,* 74, 64, 1985.
18. **Tallant, M. J., Frantz, S. W., and Ballantyne, B.,** Evaluation of the *in vitro* skin penetration potential of UCARE® Polymer JR400 using rat, mouse, rabbit, guinea pig and human skin, *Toxicologist,* 9, 61, 1989.
19. **Spencer, T. S., Hill, J. A., Feldmann, R. J. and Maibach, H. I.,** Evaporation of diethyltoluamide from human skin *in vivo* and *in vitro, J. Invest. Dermatol.,* 72, 317, 1979.
20. **Hawkins, G. S. and Reifenrath, W. G.,** Development of an *in vitro* model for determining the fate of chemicals applied to the skin, *Fund. Appl. Toxicol.,* 4, S133, 1984.
21. **Reifenrath, W. G. and Robinson, P. B.,** In vitro skin evaporation and penetration characteristics of mosquito repellents, *J. Pharm. Sci.,* 71, 1014, 1982.
22. **Hawkins, G. S. and Reifenrath, W. G.,** Influence of skin source, penetration cell fluid, and partition coefficient on *in vitro* skin penetration, *J. Pharm. Sci.,* 75, 378, 1986.
23. **Loomis, T.,** Skin as a portal of entry, in *Current Concepts in Cutaneous Toxicity,* Simon, G. A., Paster, Z., Klingberg, M. A., and Kay, M., Eds., S. Karger, Basel, 1980, 158.
24. **Stott, W. T., Dryzga, M. D., and Ramsey, J. C.,** Blood flow distribution in the $B_6C_3F_1$ mouse and the Fischer 344 rat, *J. Appl. Physiol.,* 3, 310, 1983.
25. **Frantz, S. W., Dittenber, D. A., Eisenbrandt, D. L., and Watanabe, P. G.,** Evaluation of an *in vitro* cutaneous penetration chamber using acetone-deposited organic solids, *J. Toxicol.: Cutaneous and Ocular Toxicity,* 9(4), in press, 1990.
26. **Notari, R. E.,** *Biopharmaceutics and Clinical Pharmacokinetics,* 3rd ed., Dekker, New York, 1980, 71.
27. **Ando, Y. W., Ho, N. F. H., and Higuchi, W. I.,** Skin as an active metabolizing barrier. I. Theoretical analysis of topical bioavailability, *J. Pharm. Sci.,* 66, 1525, 1977.
28. **Ando, Y. W., Ho, N. F. H., and Higuchi, W. I.,** *In vitro* estimates of topical bioavailability, *J. Pharm. Sci.,* 66, 755, 1977.
29. **Bickers, D. R. and Kappas, A.,** The skin as a site of chemical metabolism, in *Extrahepatic Metabolism of Drugs and Foreign Compounds,* Gram, T. E., Ed., Spectrum, Jamaica, NY, 1980, 295.
30. **Bickers, D. R.,** Drug, carcinogen, and steroid hormone metabolism in skin, in *Biochemistry and Physiology of the Skin,* Goldsmith, L. A., Ed., Oxford University Press, New York, 1983, 1169.
31. **Hadgraft, J.,** Theoretical aspects of metabolism in the epidermis, *Int. J. Pharm.,* 4, 229, 1980.
32. **Kao, J., Patterson, F. K., and Hall, J.,** Skin penetration and metabolism of topically applied chemicals in six mammalian species, including man: an *in vitro* study with benzo[a]pyrene and testosterone, *Toxicol. Appl. Pharmacol.,* 81, 502, 1985.
33. **Mukhtar, H. and Bickers, D. R.,** Drug metabolism in skin, *Drug Metab. Dispos.,* 9, 311, 1981.
34. **Smith, L. H. and Holland, J. M.,** Interaction between benzo[a]pyrene and mouse skin in organ culture, *Toxicology,* 21, 47, 1981.
35. **Nakashima, E., Noonan, P. K., and Benet, L. Z.,** Transdermal bioavailability and first-pass skin metabolism: a preliminary evaluation with nitroglycerin, *J. Pharmacokin. Biopharm.,* 15, 423, 1987.
36. **Santus, G., Watari, N., Hinz, R. S., Benet, L. Z., and Guy, R. H.,** Cutaneous metabolism of transdermally delivered nitroglycerin *in vitro,* in *Skin Pharmacokinetics, Pharmacology and the Skin,* Vol. 1, Shroot, B. and Schaefer, H., Eds., Karger, Basel, 1987, 240.
37. **Sheard, C., Roth, G. M., and Horton, B. T.,** Relative roles of extremities in body heat dissipation on normal circulation and peripheral vascular disease, *Arch. Phys. Ther.,* 20, 133, 1939.

38. **Skelly, J. P., Shah, V. P., Maibach, H. I., Guy, R. H., Wester, R. C., Flynn, G., and Yacobi, A.,** FDA and AAPS report of the workshop on principles and practices of *in vitro* percutaneous penetration studies: relevance to bioavailability and bioequivalence, *Pharmaceut. Res.*, 4, 265, 1987.

39. **Fritsch, W. C. and Stoughton, R. B.,** The effect of temperature and humidity on the penetration of ^{14}C acetylsalicylic acid in excised skin, *J. Invest. Dermatol.*, 41, 307, 1963.

40. **Stoughton, R. B.,** Percutaneous absorption; Influence of temperature and hydration, *Arch. Environ. Health*, 11, 551, 1965.

41. **McCreesh, A. H. and Steinberg, M.,** Skin irritation testing in animals, in *Dermatotoxicology*, 2nd ed., Marzulli, F. N. and Maibach, H. I., Eds., Hemisphere, Washington, 1983, 155.

42. **Blank, I. H., Scheuplein, R. J., and MacFarlane, D. J.,** Mechanism of percutaneous absorption. III. The effect of temperature on the transport of non-electrolytes across the skin, *J. Invest. Dermatol.*, 49(6), 582, 1967.

43. **Feldmann, R. J. and Maibach, H. I.,** Regional variation in percutaneous penetration of ^{14}C cortisol in man, *J. Invest. Dermatol.*, 48, 181, 1967.

44. **Wang, Y., et al.,** Effects of temperature and humidity on the absorption of pesticides in vitro, *Occup. Medicine*, 3, 2, 1983.

45. **Reifenrath, W. G. and Hawkins, G. S.,** The weanling Yorkshire pig as an animal model for measuring percutaneous penetration, in *Swine in Biomedical Research*, Vol. 1, Tumbleson, M. E., Ed., Plenum, New York, 1986, 679.

46. **Harrison, S. M., Barry, B. W., and Dugard, R. H.,** Factors controlling liquid and vapor diffusion through human skin for the model compound benzyl alcohol, *J. Pharm. Pharmacol.*, 35 (Suppl.), 32 P, 1983.

Chapter 4

PREPARATION OF BIOLOGICAL MEMBRANES

Robert L. Bronaugh

TABLE OF CONTENTS

I. LIVING OR NONLIVING SKIN

For a number of years *in vitro* absorption studies have been performed by using nonliving skin in diffusion cells, often with normal saline as the receptor fluid. The term diffusion cell indicates exactly the type of process being measured — the passive diffusion of substrate from one side of the barrier to the other. For many compounds, accurate rates of absorption can be determined in this way. However, for some compounds that are metabolized by enzymes at a high specific activity, a more accurate picture of absorption may be obtained by maintaining the viability of skin in the diffusion cell, as discussed later in this chapter and in other chapters in this book. Knowledge of the formation of therapeutic or toxic metabolites in skin can be of great value in assessing the effect of the absorption of a compound in contact with skin.

II. HUMAN OR ANIMAL SKIN

Human skin, of course, is preferable for all permeation studies, but its use is not always possible because of limited availability. If one is willing to use only freshly obtained viable human skin (surgical specimens), this further limits the supply since the use of cadaver skin is excluded. Therefore, the use of animal skin usually becomes necessary at some point in an *in vitro* permeability study. This can best be accomplished by ''calibrating'' an animal membrane by comparing the absorption of the test compound through the skin of the animal model with that through human skin. The skin of a hairless animal is most satisfactory for the preparation of a membrane for an absorption study. As discussed below, hair interferes greatly in the preparation of split-thickness preparations of skin. The skins of the hairless guinea pig and rat are preferable to the skin of the hairless mouse because hairless mouse skin, like all mouse skin, is very thin and therefore much more permeable than human skin.

III. THE BARRIER LAYER

A membrane used in an *in vitro* study should simulate as closely as possible the barrier layer in skin. The barrier layer refers to the thickness of skin through which a compound must diffuse *in vivo* before being taken up by blood vessels in the upper papillary dermis, and then entering the systemic circulation. The thickness of the barrier layer, which includes the whole epidermis and a small portion of the dermal tissue, certainly varies depending on the type of skin used, but is probably in the range of 100 to 200 μm.

If additional dermal tissue is present on the skin membrane used in a diffusion experiment, the effects of this tissue on absorption depend on the solubility of the chemical. A water-soluble substance will diffuse readily through the aqueous dermal tissue, and its absorption will be affected only miminally by the presence of additional tissue. However, a hydrophobic compound will diffuse through this tissue very slowly and, therefore, will appear to be absorbed much more slowly than in an *in vivo* study.

IV. FULL OR SPLIT-THICKNESS SKIN

The use of full-thickness skin is really justifiable only when the animal skin is already very thin, such as that of the mouse (400 μm)[1] or rabbit. For other animals, such as the rat (800 to 870 μm),[2] guinea pig, monkey, and pig, full-thickness skin is almost 1 mm thick, and for human skin it can be several millimeters thick.[3] Therefore, some means should be used to prepare a membrane that accurately reflects the barrier layer in thickness. As mentioned above, this is particularly important when a hydrophobic compound is examined.

V. PREPARATION OF SPLIT-THICKNESS SKIN

A. DERMATOME

The use of the dermatome is the best way of preparing biological membranes for percutaneous absorption studies. Unlike other methods which are described in this chapter, a dermatome can be used with hairless or haired skin, and it can be used without adversely affecting the viability of the membrane. For the last 6 years we have prepared membranes almost exclusively with a dermatome—[4-5] the Padgett Electrodermatome (Padgett Instruments, Kansas City, MO).

Full-thickness skin from a human subject or an animal is pinned (epidermis side up) to the surface of a block for cutting. A styrofoam block is convenient because it can be readily shaped to the desired size by cutting with a knife and because pins for anchoring the skin can be easily punched into this material. The width of the styrofoam block must be less than the length of the cutting edge of the dermatome blade so that the blade can contact the surface of the skin. The piece of skin should overlap the edges of the block so that it can be attached with the pins to the sides of the block (so that the pins are out of the way of the dermatome blade).

The depth of the cut is controlled with a lever on the side of the dermatome head, with the indicated calibrations being in thousandths of an inch. The actual thickness of the membrane obtained is a result of the pressure applied and the angle at which the dermatome is held as it is pushed across the skin. We have found it helpful to check the thickness of each membrane prepared with the dermatome by using a micrometer (Mitutoyo micrometer, 0.01 to 9 mm, L. A. Benson Inc., Baltimore, MD). With a little practice it is possible to prepare skin sections of reproducible thickness.

B. SEPARATION AT THE DERMAL/EPIDERMAL JUNCTION
1. Heat Separation

Elevation of temperature has been used for a number of years to loosen the bond between the epidermis and dermis for the preparation of epidermal sheets. Baumberger et al.[6] placed full-thickness human skin on a hot plate for 2 min at a temperature of 50°C and found that the epidermis could easily be removed from the dermis by using blunt dissection. Heated water is used more commonly for this separation because of better temperature control, which leads to more reproducible results.[7,8] Water is heated in a beaker to 60°C. Full-thickness human skin (with underlying muscle and fascia removed) is suspended in the water with forceps for 30 s. The epidermis is then removed from the dermis by using foceps to gently pull the epidermis away. The ease of separation varies somewhat from specimen to specimen. The dissection can be facilitated by pinning the skin to a block so that it is stretched tight.

Heat separation of skin is useless, however, for preparing a membrane for absorption studies with haired skin. The hair shafts remain in the dermis, causing holes to be created in the epidermal membrane as it is pulled from the dermis. Therefore, only the skin from hairless animals can be separated by this procedure. A different length of exposure to heat may be required for separation of animal skin.

The effect of heat on the viability of skin is of concern in absorption/metabolism studies. The enzymatic hydrolysis of diisopropyl fluorophosphate in human skin appeared to be inactivated during the heat separation procedure.[3] However, many enzymes may be only minimally affected by short exposure to a temperature of 60°C.

2. Chemical Separation

The epidermis and dermis have been separated after soaking full-thickness human and animal skin in different chemical solutions. Exposure times are usually measured in hours and, therefore, viability of skin is probably lost.

The effects of 2 *M* solutions of various salt anions and cations on the separation of the epidermis have been investigated with human skin.[9] The separation of the epidermis was achieved by acids and bases at pH values that caused swelling of the collagen. The most potent anions were bromide, thiocyanate, and iodide ions, whereas acetate, sulfate, and citrate ions were ineffective in causing separation.

The advantage of chemical separation is that it appears to be effective in separating the epidermis from the dermis of haired animals under certain conditions. The barrier properties remain unaltered because the hair shaft comes out of the dermis during separation and stays in place in the epidermis. Scott and co-workers[10] reported that after skin of 28 day-old Wistar rats was soaked in 2 *M* sodium bromide for 24 h, epidermal membranes could be separated from the dermis. However, this procedure was ineffective with skin of older rats (7-8 weeks of age).

3. Enzyme Separation

A few studies have been reported in the literature concerning the separation of the epidermis and dermis by incubation of skin in enzyme preparations. The protease dispase produced an epidermal sheet that could be easily peeled from the dermis of human skin after a 24-h incubation at 4°C.[11] A crude bacterial collagenase was effective in the preparation of epidermal sheets at concentrations of 0.1 and 0.2% after a 3-h incubation at 37°C.[12] Unlike the cells produced in the dispase separation, however, most of the cells were found to be nonviable following separation, but differences in the temperature of the incubations may have been responsible. The use of the enzymes pancreatin and trypsin for epidermal-dermal separation has also been described.[13] The referenced articles should be consulted for further details.

VI. STORAGE OF MEMBRANES

Occasionally it may be desirable to store a portion of a valuable membrane (such as human skin) for use at a later time. This should be done only when passive diffusion is to be measured, since techniques for storing membranes for permeability measurements have not been examined for maintenance of viability.

In studies involving the use of human cadaver skin for diffusion measurements, we examined the use of a standard compound (tritiated water) for assessing the integrity of the barrier layer.[14] A barrier check should be done even with newly obtained cadaver skin, and it should also be repeated following storage. The ^{3}H-water permeability of human cadaver skin was measured initially, and then the skin was stored in an air-tight plastic freezer bag at -20°C. After various time periods, the skin was thawed and the barrier properties were rechecked with ^{3}H-water (Table 1). Significant increases in water permeability were seen with the skin of two donors (Nos. 39 and 40) after 2 months of storage. In most cases of storage ranging from 2 to 12 months, water permeability did not change, indicating a maintenance of the barrier properties.

^{3}H-water permeation has often been measured by applying an excess of compound to the surface of the skin and measuring the steady-state rate of permeation, e.g., by taking at least 4 to 5 measurements at hourly intervals following the application of ^{3}H-water. A permeability constant (K_p) can be determined by dividing this rate by the initial concentration of applied material.

We observed that a much more rapid evaluation of water permeability could be made in what we called the 20-min test. By using diffusion cells with an exposed skin area of 0.32 cm^2, 100 μl of ^{3}H-water (approximately 0.3 μCi) was applied to the surface of the skin so that it was completely covered. The top of the cells was occluded with Parafilm. After 20 min, the unabsorbed material was blotted from the surface of the skin with a cotton-

TABLE 1
Effect of Length of Frozen Storage on Water Permeation[a]

Skin donor no.	Initial $K_p \times 10^3$ (cm/h)	Final $K_p \times 10^3$ (cm/h)	Length of storage (months)
34	0.98 ± 0.13	0.84 ± 0.13	2
39	1.98 ± 0.08	3.48 ± 0.58[b]	2
40	1.22 ± 0.28	2.02 ± 0.05[b]	2
41	1.43 ± 0.15	1.57 ± 0.06	2
36	1.05 ± 0.19	1.36 ± 0.22	5
26	0.93 ± 0.03	1.35 ± 0.23	12
29	2.13 ± 0.09	1.89 ± 0.10	12
30	1.81 ± 0.10	2.24 ± 0.26	12
32	0.97 ± 0.08	1.06 ± 0.07	12

[a] Values are the mean ± SE of 3 to 4 determinations.
[b] Statistically different from initial K_p value by Student's *t*-test, $p < 0.05$.

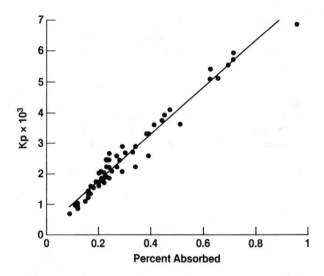

FIGURE 1.   Correlation of results from two methods of measuring water permeation through skin: permeability constant (K_p) and percent dose absorbed (20-min test).

tipped applicator, and then the surface of the skin was rinsed once with distilled water. Effluent from the flow cell was collected for an additional 60 min, at which time radioactivity in the effluent had returned to the baseline level. ^{3}H-water absorption was expressed as the percent of the applied dose absorbed. A good correlation (r = 0.98, $p < 0.01$) was obtained between the values measured in the 20-min test and the corresponding permeability constants obtained from the same membranes (Figure 1).

The ability to store skin satisfactorily in the frozen state has been the subject of controversy. Some investigators have found that frozen human and animal skin can be stored without deterioration of the barrier.[15-17] Hawkins and Reifenrath[18] observed, however, that pig skin stored frozen longer than 1 week became increasingly permeable to *N,N*-diethyl-*m*-toluamide. The penetration of T-2 toxin was shown to increase through human and monkey skin that had been stored at −60°C.[19] It seems advisable to check the integrity of the skin barrier by using either a standard compound or the test compound following frozen storage. It has been demonstrated that damage to the barrier properties can sometimes occur. It should be remembered that the viability of skin may also be altered by frozen storage techniques.

REFERENCES

1. **Behl, C. R., Flynn, G. L., Kurihara, T., Smith, W. M., Bellantone, N. H., Gataitan, O., and Higuchi, W. I.,** Age and anatomical site influences on alkanol permeation of skin of the male hairless mouse, *J. Soc. Cosmet. Chem.,* 35, 237, 1984.
2. **Yang, J. J., Roy, T. A., and Mackerer, C. R.,** Percutaneous absorption of benzo(a)pyrene in the rat: comparison of in vivo and in vitro results, *Toxicol. Ind. Health,* 2, 409, 1986.
3. **Loden, M.,** The in vitro hydrolysis of diisopropyl fluorophosphate during penetration through human full-thickness skin and isolated epidermis, *J. Invest. Dermatol.,* 85, 335, 1985.
4. **Bronaugh, R. L. and Stewart, R. F.,** Methods for *in vitro* percutaneous absorption studies. III. Hydrophobic compounds, *J. Pharm. Sci.,* 73, 1255, 1984.
5. **Bronaugh, R. L. and Stewart, R. F.,** Methods for *in vitro* percutaneous absorption studies. VI. Preparation of the barrier layer, *J. Pharm. Sci.,* 75, 487, 1986.
6. **Baumberger, P. J., Suntzeff, V., and Cowdry, E. V.,** Methods for the separation of epidermis from dermis and some physiologic and chemical properties of isolated epidermis, *JNCI,* 2, 413, 1942.
7. **Scheuplein, R. J.,** Mechanism of percutaneous absorption. I. Routes of penetration and influence of solubility, *J. Invest. Dermatol.,* 45, 334, 1965.
8. **Bronaugh, R. L., Congdon, E. R., and Scheuplein, R. J.,** The effect of cosmetic vehicles on the penetration of N-nitrosodiethanolamine through excised human skin, *J. Invest. Dermatol.,* 76, 94, 1981.
9. **Felsher, Z.,** Studies on the adherence of the epidermis to the corium, *J. Invest. Dermatol.,* 8, 35, 1947.
10. **Scott, R. C., Walker, M., and Dugard, P. H.,** *In vitro* percutaneous absorption experiments: a technique for the production of intact epidermal membranes from rat skin, *J. Soc. Cosmet. Chem.,* 37, 35, 1986.
11. **Kitano, Y. and Okada, N.,** Separation of the epidermal sheet by dispase, *Br. J. Dermatol.,* 108, 555, 1983.
12. **Hentzer, B. and Kobayasi, T.,** Enzymatic liberation of viable cells of human skin, *Acta Dermatol. Venereol.,* 58, 197, 1978.
13. **Omar, A. and Krebs, A.,** An analysis of pancreatic enzymes used in epidermal separation, *Arch. Dermatol. Res.,* 253, 203, 1975.
14. **Bronaugh, R. L., Stewart, R. F., and Simon, M.,** Methods for *in vitro* percutaneous absorption studies. VII. Use of excised human skin, *J. Pharm. Sci.,* 75, 1094, 1986.
15. **Harrison, S. M., Barry, B. W., and Dugard, P. H.,** Effects of freezing on human skin permeability, *J. Pharm. Pharmacol.,* 36, 261, 1984.
16. **Franz, T. J.,** Percutaneous absorption. On the relevance of *in vitro* data, *J. Invest. Dermatol.,* 64, 190, 1975.
17. **Del Terzo, S., Behl, C. R., Nash, R. A., Bellantone, N. H., and Malick, A. W.,** Evaluation of the nude rat as a model: effects of short-term freezing and alkyl chain length on the permeabilities of n-alkanols and water, *J. Soc. Cosmet. Chem.,* 37, 297, 1986.
18. **Hawkins, G. S. and Reifenrath, W. G.,** Development of an *in vitro* model for determining the fate of chemicals applied to skin, *Fund. Appl. Toxicol.,* 4, S133, 1984.
19. **Kemppainen, B. W., Riley, R. T., Pace, J. C., and Hoerr, F. J.,** Effects of skin storage conditions and concentration of applied dose on [³H]T-2 toxin penetration through excised human and monkey skin, *Food Chem. Toxicol.,* 24, 221, 1986.

Chapter 5

METHODOLOGY FOR THE EXECUTION OF IN VITRO SKIN PENETRATION DETERMINATIONS

George S. Hawkins, Jr.

TABLE OF CONTENTS

I. INTRODUCTION

A variety of experimental methods are used *in vitro* to increase our understanding of the process of skin penetration. The experimental methods described are primarily intended to provide estimates of *in vivo* skin penetration. To achieve this, the innate permeability of skin must be maintained *in vitro*. For some compounds the skin viability[1] may also be important. Both of these characteristics can be preserved by using the appropriate receptor fluid and conditions.[2]

Another important function of the receptor fluid is the capacity of the fluid to accept compounds which have penetrated through skin. *In vivo,* compounds which penetrate through skin are generally removed by the microcirculation which effectively creates a sink condition. *In vitro,* the receptor fluid should also provide a sink condition.

Due to the normal variation in skin permeability,[3] the penetration values from *in vitro* skin penetration determinations can be highly variable. To account for this, a number of replicates should be made with different skin samples for any particular determination. The appropriate number of replicates can be determined using an analysis of statistical power.[4]

Results may be expressed differently depending on the type of determination. For example, percent absorption is used in finite dose experiments, whereas steady state flux is frequently used for infinite dose determinations. Comparisons between results from different studies are sometimes made to validate *in vitro* methods or the use of a particular skin. If comparisons are made between penetration values, they should include a number of compounds at equivalent doses. Sets of values are often compared by correlation or rank order. A comparison with human *in vivo* values can be made for some compounds, using data from the literature.[5-7]

The determination of skin penetration for lipophilic compounds presents special difficulties. *In vivo,* human and pig penetration values for a number of compounds were comparable. However, when penetration values using full thickness pig skin[8] *in vitro* were compared to the human *in vivo* values, the penetration values for hydrophilic compounds were similar but the *in vitro* values for lipophilic compounds underestimated *in vivo* absorption. The use of split thickness skin improved the agreement while a further improvement was gained by adding the dermal residue to the recovery from the receptor fluid[9] (Figure 1). For lipophilic compounds, dermal retention occurring *in vitro* but not *in vivo* was largely the cause of the discrepancy.[10] For volatile compounds, air flow was also found to influence the agreement of values (see Section VII). In addition, differences between *in vitro* and *in vivo* values for lipophilic compounds may occur when the *in vivo* method is based on excretion since compounds or metabolites may be excreted over a time interval which exceeds the duration of the experiment.

II. EQUIPMENT SELECTION

Selecting equipment for *in vitro* skin penetration determinations depends on individual requirements. Most of the laboratory equipment can be obtained from commercial sources. A generalized list with sources is provided in the appendix. The skin penetration cell should be constructed with a material which will not retain or react with the test compounds.

Experiments may be conducted with air on both sides of the membrane, liquid on both sides, or air on the epidermal side and liquid in contact with the dermal side. Additionally, the air and liquid may remain static or flowing during determinations. Various cell designs are available to suit these options. Side-by-side cells serve as a donor-acceptor pair with the skin held vertically between the cells. In upright designs, the skin is mounted horizontally over the mouth of a lower cell. An upper cell may be attached to allow for the recovery of volatile compounds (Figure 2). It may also be used to control the air temperature, flow, and humidity.

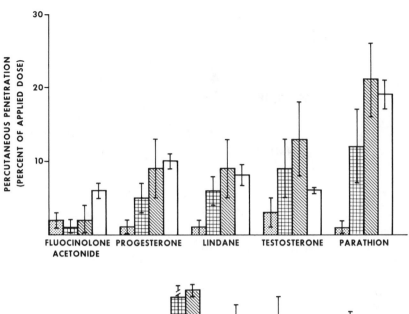

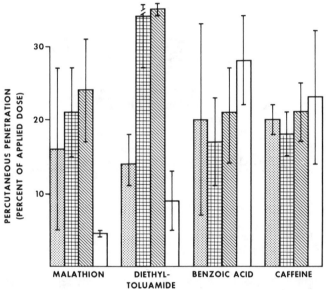

FIGURE 1.   Pig skin penetration values obtained from full thickness skin (::), split thickness skin (+ +), and split thickness skin plus dermal recovery (\\ ) vs. *in vivo* ( ). (Reproduced with permission of the copyright owner, Plenum Press).

Dosing methods should also be considered in the selection of skin penetration cells. For example, it would not be practical to apply a low viscosity fluid to the surface of the skin in a side-by-side system since it would flow off the surface. However, a side-by-side design is more appropriate if both cells will be filled with liquid.

Bubbles often form under the skin when it is horizontally mounted. They should not be allowed to accumulate under the surface of the skin since they reduce the area of perfusion and therefore may reduce the measured penetration. One way to diminish the impact of bubble formation is by using a cell which permits a larger piece of skin since the effect of a bubble of a given size will be less with increasing skin area. Bubble traps are also available to scavenge bubbles in flow through systems.

The distinction between the flow and static penetration cell design is somewhat arbitrary

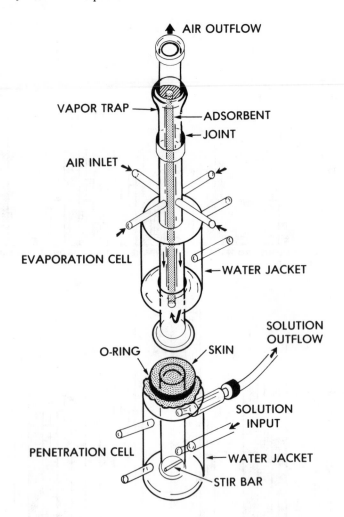

FIGURE 2.   Skin penetration cell, evaporation cell, and vapor trap. (Reproduced with permission of the copyright owner, *J. Pharm. Sci.*).

since a static system with a sampling port can be sampled by replacing all the fluid each time so that a time course of penetration can be made. However, this may cause air bubbles under the skin and can be impractical when numerous cells are in operation. Sample collection from numerous cells is much easier using flow-through cells and a fraction collector. Time course of penetration information from replicate cells is also useful in identifying abnormal penetration patterns from cells with skin which has uncharacteristic permeability.

Flow systems also require a multichannel peristaltic pump to maintain flow from a reservoir to the penetration cells. A one quart cubitainer or i.v. bag can be used as an isolated reservoir. Alternatively, a stoppered glass vessel vented to the outside with sterile cotton may be used. A glass manifold to connect tygon tubing from the reservoir to each penetration cell is available. A plastic bulkhead fitting can be used to attach the manifold to the cubitainer. The outflow from the penetration cells can be routed to a multichannel fraction collector or large fractions can be collected manually in standard glassware. A static cell design may be necessary to achieve detectable levels with compounds which have a minimal penetration. However, the effect of a limited supply of oxygen, nutrients and buffering to support viable skin should be considered.

III. RECEPTOR FLUID COMPOSITION AND FLOW

In vivo, the living cells of the skin maintain the integrity of the tissue. These cells also may interact with a penetrating compound by rendering it more water soluble or altering the toxicity[1] of the compound. Once removed from the donor organism, the skin cells will die and the tissue will degrade unless the means to maintain viability are provided. Therefore, *in vitro,* the receptor fluid should not only act as an acceptor for penetrating compounds, but should provide the water, biochemicals and ions needed at the proper pH (7.2 to 7.3) and osmotic strength. This is particularly important with slowly penetrating compounds in which a determination is conducted for extended periods and in cases where the penetrant may be metabolized.

The capacity of the receptor fluid to accept compounds from the skin primarily depends on the solubility of the penetrant in the fluid and the rate of flow. If a compound is insoluble, then the transfer of penetrant into the media will be minimal and the apparent penetration value inaccurate. Depending on the rate of penetration, this can be a problem for determinations involving lipophilic compounds. Attempts to solve this problem have been made by using nonionic surfactants[11] or serum. Serum would seem to be the most appropriate fluid to maintain viability and obtain representative skin penetration values. However, with full thickness pig skin, substituting serum for a salt solution did not improve the agreement with *in vivo* values.[8] It is possible that the full thickness dermis may have prevented any improvement. In some cases, the agreement has been improved with serum.[12]

For many determinations, commercial media may be suitable. Variants of Eagle's minimal essential media are available which have been shown to support skin viability.[2,13] In addition, they offer the advantages of standard composition and convenient packaging.

When using a flow through system, the flow of fluid through the penetration cell should be adjusted in regard to the applied dose, compound solubility and skin surface-to-cell volume ratio. For compounds which have a significant penetration rate, larger applied doses are associated with greater mass transfer. In this case, greater flow through the penetration cell may be needed. If the hourly samples are to be used to estimate hourly penetration, a flow of at least one penetration cell volume per hour should be used. When cells with small volumes relative to skin surface area are used, one cell volume per hour would probably be insufficient to obtain effective clearance. For penetration determinations with highly lipophilic compounds, a comparison of results using two flow rates can be done to determine if the penetration rate is limited by the flow rate (Figure 3). In this case no differences were evident. If greater penetration rates occur using greater flows, then the flow rate should be increased to eliminate this effect.

IV. TEMPERATURE AND HUMIDITY

The process of skin penetration is based on diffusion and is therefore influenced by the temperature of the skin. *In vivo,* the surface of the skin is exposed to a wide range of temperatures while the interior is closer to the core temperature. To provide a similar interior condition *in vitro* the receptor fluid temperature can be controlled. Water jacketed penetration cells are available for this purpose.

Exposing the surface of the skin *in vitro* to variable temperatures is not desirable since this factor has been shown to affect the rate of skin penetration.[8,14] Therefore, it is also recommended that the skin surface be exposed to a controlled environment. A laboratory with a well controlled ventilation system may be sufficient. As an alternative, evaporation cells can be installed which receive incoming air from a water jacketed coil connected to a circulating water bath. An additional level of control can be gained using water jacketed evaporation cells.

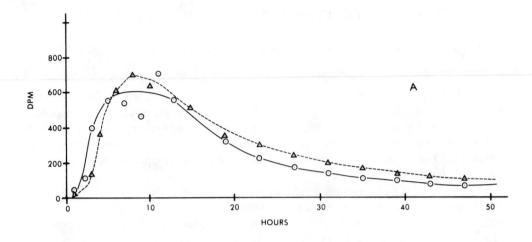

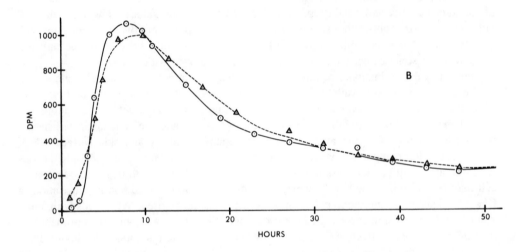

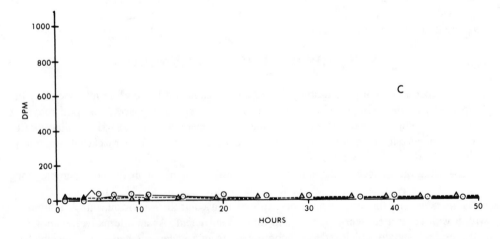

FIGURE 3.   Percutaneous penetration of radiolabelled compounds N-N, diethyltoluamide (A), caffeine (B), and lindane (C) at 5 ml (- - - - -) vs. 10 ml/h (-------) using cells of 5 ml capacity. (Reproduced with permission of the copyright owner, *Fund. Appl. Toxicol.*).

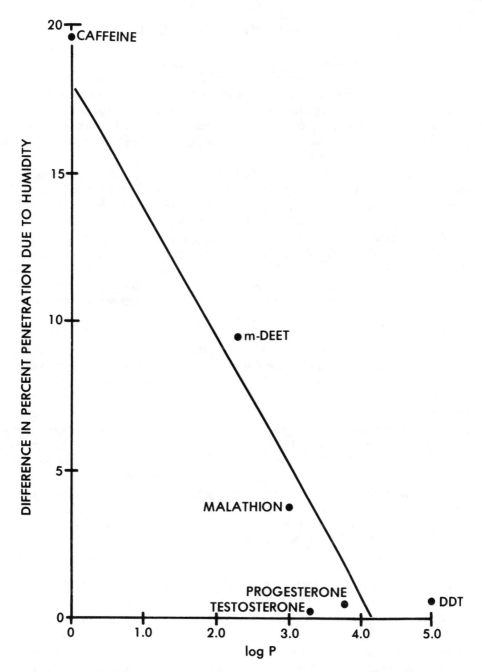

FIGURE 4.   Difference in percentage percutaneous penetration vs. lipid solubility (Log P, logarithm of the octanol to water partition coefficient) at zero and high (approximately 75%) relative humidity. (Reproduced with permission of the copyright owner, *Fund. Appl. Toxicol.*).

High levels of humidity have been shown to increase skin penetration *in vivo*.[14] *In vitro,* the rate of skin penetration for more water soluble compounds was also greater with higher humidity[8] (Figure 4). This is thought to be the result of increased hydration of the stratum corneum. Therefore, the relative humidity of the laboratory should be monitored. If the relative humidity is extremely high or variable it can be standardized by passing the air through a drierite column before it enters the evaporation cell.

V. SKIN PROCUREMENT

When *in vitro* skin penetration experiments are conducted to model human *in vivo* skin penetration, obtaining suitable skin can be difficult. Human abdominal skin is a good choice since it has an intermediate permeability[3] and may be obtained from surgery or postmortem. Unfortunately, few laboratories have a readily available source. Consequently, *in vitro* skin penetration data from a variety of animal skins are reported in the literature. Comparing the permeability of different animal skins to that of man is difficult since the methods and equipment are often dissimilar.

If human skin is available it should be held on gauze dampened with receptor fluid to prevent dehydration prior to use. If it is not used directly after excision it can be held overnight in a refrigerator. To prepare the skin for *in vitro* use, the subcutaneous fat should be removed by scraping the fat off with a scalpel or with a dermatome as described below.

If human skin is unavailable, some other source must be chosen. The objective of the determinations will limit the choice. One suitable alternative is split thickness (1.0 mm) skin from weanling Yorkshire pigs.[9] Pig skin has been used because the morphology and permeability are reasonably similar to human skin.[7] Also, sufficient quantities can be procured as needed for a number of cells and experiments. Female pigs are preferred since urine recovery is more practical for any subsequent *in vivo* work.

To obtain pig skin for *in vitro* use, a weanling pig is sacrificed and the hair removed from the skin on the dorsal sides of the thorax using electric clippers. A 30 × 40 cm rectangle is cut 5.0 mm deep into each side with a scalpel. The skin is then grasped by one end of each rectangle and pulled free from the underlying muscle. Smaller skin strips are cut from this piece and placed on a board over a 3-cm thick layer of 10 × 20 cm gauze damped with receptor fluid. The gauze and skin are secured to the proximal end of the board using sheet metal pliers or some other suitable clamping device. A dermatome set at 1.0 mm is positioned on the proximal end of the skin strip just in front of the pliers or clamp. The dermatome is then pushed across the epidermal side of the skin while the top 1.0 mm of skin is pulled from the opening of the dermatome by an assistant with a forceps. The thickness of the skin should be checked using a caliper. The skin should be cut into an appropriate size for the penetration cells and mounted as soon as possible.

Irrespective of source, skin should be free from disease and damage such as sunburn or abrasion. It should be carefully handled and received from the donor in a timely manner. Skin should not be stored frozen prior to use unless first checked for changes in permeability.[8]

VI. DOSE SELECTION, FORMULATION, AND APPLICATION

The dose, formulation and application of the penetrant can affect the results of a determination.[15] The dose may be far in excess of what can pass through the skin during an experiment (infinite) or conversely, be depleted during the experiment (finite). The dose and application method are often selected to mimic an *in vivo* circumstance. A single finite dose might be chosen to represent a single exposure to a material such as a pesticide or mosquito repellent. An infinite dose could be applied to model exposures from skin therapy, transdermal delivery of systemic drugs or extreme toxin exposure. In determinations where the penetration cell medium is to be analyzed for metabolic alteration of the applied compound, a range of doses may be required to determine the capacity of the skin to transform the dose.

Compounds may be applied neat or in a formulation. The compound may be formulated alone in a solvent or in a complex mixture. The stability and compatibility of the formulation should be evaluated before use. Solids or high viscosity formulations can be applied on a weight basis using a glass or teflon rod as an applicator. Lower viscosity liquids are applied by volume using a syringe. Finite doses should be formulated in a volume which is suitable for the surface area of application. If too great a volume is applied, liquids will concentrate around the edge of the evaporation cells (if present) or else flow beyond the application area. Conversely, an inadequate volume will not cover the desired area.

Care should be exercised to avoid increases in solute concentration from evaporation which may occur when using highly volatile solvents. Small vials with teflon valves are available for withdrawal of liquids by syringe when using highly volatile or hazardous compounds.

VII. PREPARATION AND EXECUTION OF *IN VITRO* DETERMINATIONS

To assure success, *in vitro* skin penetration systems with numerous skin cells and external connections should be assembled a day prior to the execution of an experiment. This allows time to check the system for leaks and the proper operation of water baths or pumps.

A. SKIN PENETRATION

The following procedures are intended as a general guide for systems which resemble those available from Laboratory Glass Apparatus Inc. Procedures intended to preserve skin viability are included. Penetration cells should be treated with a silicone coating (Prosil 28) to reduce retention of the penetrant on surfaces. If a flow system is used with media as a receptor fluid, the tubing and reservoirs should be sterilized to prevent contamination. Sterilization using ethylene oxide is conveniently done overnight. The items to be sterilized are placed inside a plastic bag with an ampule of ethylene oxide. The bag is closed tightly and the ampule broken open. The following morning the bag is opened in a hood and the items removed for assembly. Alternatively, a 10% dilution of commercially available laundry bleach can be used to decontaminate the reservoir and lines followed by thorough rinsing. Glass penetration cells may also be sterilized by autoclaving.

Media should be prepared a day in advance in a clean volumetric flask. With flow systems, the volume of media prepared is based on the flow rate, duration of the determination and number of skin penetration cells. The addition of an antibiotic is recommended to prevent microbial growth. Gentimicin has been effective at 0.05 g/l.[16] The pH is checked, adjusted if needed and the media is sterile filtered with a 0.45 μ filter.

For reservoirs such as cubitainers which are isolated from the outside, the dissolved oxygen in the media can be increased to promote viability. To accomplish this, an open cubitainer filled with media is degassed in a vacuum dessicator for approximately ten minutes using a vacuum pump with an in line cold trap. The dessicator is then vented with bottled 95% oxygen-5% carbon dioxide. A sterile Pasteur pipet attached to tubing from the cylinder is inserted into the media. The media is regassed by bubbling for approximately 5 min. The cubitainer or other reservoir is then connected to the tubing which leads to the peristaltic pump and penetration cells.

Prior to installing the skin on the penetration cells, the water baths should be operating and the media flowing to each cell. When the cells are full, the flow is stopped and the skin installed. The peristaltic pump is restarted and air trapped under horizontally mounted skin is removed via the outflow tubing by tilting the cell.

The dose should be applied to the skin in a timely fashion so that any loss of volatile compounds is minimal and the first samples represent nearly equivalent intervals. If a fraction collector is used, sample collection should begin when the last skin sample is dosed. During collection the flow rate can be checked using a calibrated test tube. Following dose application, control vials should be prepared to establish a mean dose.

Difficulty in obtaining equivalent flow rates to all penetration cells may occur when a cassette pump is first started. To attain equivalent flows, the pressure on the tubing inside each cassette should be checked during the experiment. Once the tubing has been used for several hours there should be less than 10% difference between the flow rates. If the flow rate is sufficient to maintain sink conditions, small variations in the flow rate are unlikely to affect the penetration (Figure 3).[8]

Experiments are often conducted for 24 to 48 h to allow the material applied to the skin to reach a peak penetration rate or, in the case of an infinite dose, a constant flux. Compounds with appreciable penetration flux applied in the 4 to 1000 $\mu g/cm^2$ range generally can be expected to reach a peak penetration rate and decrease to some minimal level after 24 h. Volatile compounds usually reach a minimal evaporation rate by this time as well. Longer determinations may be required in some cases. Data from the first experiment in a series can be used to adjust the system if necessary.

B. RECOVERY OF VOLATILE COMPOUNDS

Evaporation cells should be silicone coated and connected to the air intake-outflow system. Flow meters should be connected to the air intake and outflow sides of the evaporation cells to monitor air flow. To maintain airflow, a rotary vane pump is connected to the flow meters on the exhaust side of the evaporation cells.

Evaporation traps should be filled with absorbant such as Tenax powder or aluminum oxide a day before the determination. Traps can be filled by first inserting a cotton plug in the lower end of the trap to retain the absorbant powder. A small plastic funnel can be inserted in the top of the trap to facilitate filling. The powder should not be packed tightly as this may restrict the air flow. Another cotton plug is then put over the column of powder inside the trap.

The airflow can be controlled by mounting a valve on the outflow flowmeter. The system can be checked for leaks by comparing the air flow readings of the intake and outflow meters. A leak is indicated when the intake flow is less than the outflow. A flowrate should be chosen which replaces the volume of the evaporation cell often so that vapor does not accumulate and condense on the interior of the evaporation cell. Airflow may affect penetration values of volatile compounds. For example, when air flow was increased from 60 to 600 ml/min using the cells shown in Figure 2, the penetration values for the volatile compounds *N,N*-diethyl-*m*-toluamide and malathion agreed more closely with human *in vivo* penetration values.[9] If large amounts of the dose are recovered from rinsing the upper surface of the evaporation cell at the end of the experiment this could be caused by a low flow rate.

Vapor trap sampling should be done frequently at the beginning of the determination since the evaporation rate will change the most during this period of time. Samples may conveniently be taken hourly during the first eight hours after dosing. The penetration samples can be processed for analysis at the same time to prevent a backlog of work at the end of the experiment. If the determination will proceed unattended overnight, the overnight vapor trap sample from each cell can be used to estimate the midpoint rate for that interval.

Vapor traps in use are replaced for sampling by first removing the outflow fitting and inserting this fitting into the fresh vapor trap. The used trap is then removed from the evaporation cell and replaced by the fresh trap with fitting attached. The air flow is checked and the sequence repeated for the remaining cells. The powder is removed from the trap by taking the cotton plug out of the lower end of the trap with forceps while holding the end of the trap over a container. The powder can be made to flow out the tube by gently tapping the trap against the inside of the container. In the case of radiolabeled compounds, the powder can be placed directly into a scintillation vial. The inside of the trap is then flushed with a small amount (2 to 3 ml) of alcohol to complete the recovery of the sample. The vial is then filled with 10 ml of scintillation counting solution and shaken. The sample can often be adequately extracted by the counting solution and counted in the same vial after the absorbant settles to the bottom. For unlabeled compounds, the absorbant can be extracted by adding an organic solvent to the scintillation vial at the time of collection and separating the absorbant from the solvent using a scintered glass filter funnel in a vacuum beaker. The absorbant should be rinsed while in the funnel to complete the extraction. The concentration of the samples can then be adjusted for gas chromatographic or other suitable analysis.

C. CONCLUDING A DETERMINATION

Upon completion of an experiment, the disposition of the applied dose should be determined. If a radiolabeled compound was applied, the media fractions can be analyzed by scintillation counting. If unlabeled compound was used, the receptor fluid can be collected in larger fractions and concentrated with a rotary evaporator. Depending on the method of analysis, this concentrate may then be analyzed directly for the applied compound or extracted with an organic solvent to recover the applied compound. All surfaces which have been exposed to the penetrant should be rinsed for recovery and analysis including the evaporation-penetration cells. Failure to account for the entire dose creates uncertainty in the results. After the skin sample has been removed from the penetration cell, the skin should be prepared so that the dose remaining with the skin can be localized. For small cells, the skin can be separated into a center dosed portion and an outer undosed portion using a sharpened cork borer. Scissors can be used for larger samples. If possible, epidermis should be removed from the center portion. Human epidermis can sometimes be removed using forceps. The epidermis is more difficult to remove from pig skin. A microtome equipped with a carbon dioxide freezing stage can be used for pig skin. The center skin disc is placed in the center of the microtome stage. The smallest volume of water which will allow the skin to stick to the stage, (approximately 100 $\mu l/cm^2$) is delivered between the stage and the underside of the skin with a pipetor. A microscope slide can be used to flatten the skin on the stage. A flow of carbon dioxide is started from an unregulated cylinder. After approximately 10 s, the freezing should be complete. A 100 μm cut is made to remove the epidermis and a small amount of dermis. The remaining dermis can be further sliced or allowed to thaw and removed from the stage. Before the next sample is sliced, the stage should be decontaminated by blotting with a cotton ball dampened with ethanol. The microtome blade and slide, if used, should be decontaminated with another cotton ball. For radiolabeled compounds, the cotton can be placed in a scintillation vial and filled with counting solution for counting. Unlabeled compounds can be extracted from the cotton for analysis. Dermal slices may be analyzed separately to provide a depth profile or extracted together for a total dermal recovery. Tissue solubilizer can be used to recover radiolabeled compound remaining in the skin or solvent can be used to extract unlabeled compound from the skin. Extracting the skin allows the extract to be analyzed for skin metabolism by measuring proportions of parent compound and possible metabolites. Sonication can also be used to facilitate extraction. If the compound applied was labeled with ^{14}C or ^3H, a tissue burning oxidizer can be used to recover the label remaining in the dermis, epidermis and outer skin.

If a microtome with freezing attachment is not available, in some cases, other methods can be used to localize the applied dose. For example, compound remaining on the surface of the skin may be recovered with a solvent-dampened swipe and analyzed chemically or radiometrically. The stratum corneum of some skins can be removed using successive applications of cellophane tape. The tape can then be placed in counting solution or solvent for analysis. If a significant amount of the applied dose is recovered from the outer skin, the dose was not confined to the area over the opening of the penetration cell so the effective dose per unit area was less than intended.

VIII. ROBOTICS

The introduction of robotics into the laboratory has been a significant innovation. The

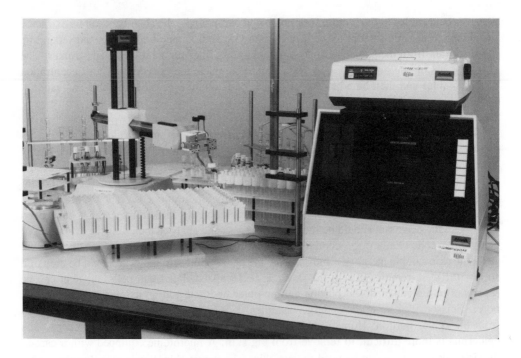

FIGURE 5.   Robotically sampled skin penetration system.

appeal is broad, as robotics offer the opportunity to reduce labor costs as well as free skilled technicians from repetitive operations. Other benefits are improved sampling precision and the ability to operate in hazardous or toxic environments.

The sampling of percutaneous penetration systems is a logical candidate for the application of robotic technology because the permeability of skin can vary greatly. This phenomenon creates a requirement for numerous replicates.

A typical robotic system consists of a controller for programming robotic movements, a mechanical arm to execute commands from the controller, modules which perform specific tasks such as weighing, capping or dispensing and work stations where tasks unique to the application are performed. In general, programming involves the assignment of locations for the robotic arm in a suitable sequence coupled with the appropriate commands to various automated work stations. Laboratory systems have been developed which represent two approaches to accessing and manipulating laboratory paraphenalia. In one solution the robotic arm operates from a fixed location central to a circular work area (Figure 5). In the other approach the robotic arm is mobile and operates along a track. The application may dictate the choice.

The decision to use robotics should be based on the individual requirements of the application. For example, if there is no requirement for the determination of evaporation profiles of volatiles, the sampling task is greatly simplified. The gains should be compared to the costs of maintenance and implementation as well as the interval during which any given application is likely to be required. The use of robotics is most appropriate when the procedures are easily accomplished by the robotic system (failure rate approaches zero) and the amount of labor saved is large.

APPENDIX

Item	Source
Skin Permeation System, (model LG-1083)	Laboratory Glass Apparatus, Inc., Berkeley, CA
Media	Grand Island Biological, NY
Rotary vane vacuum pump (Gast), Filter apparatus (Deltaware 47 mm for media sterilization), Cubitainer (1 l), and Tygon Tubing (R-3603)	VWR, Inc., San Francisco, CA
Peristaltic Pump (model 7620-52)	Cole Parmer, Instrument, Co., Chicago, IL
Gentimicin Sulfate (no. G-7507)	Sigma Chemical Co., St. Louis, MO
Ethylene oxide (Anaproline AN 73)	H. W. Anderson Products, Oyster Bay, NY
Disposable microtome blades (Tissue Tek #4689) and Silicone glassware coating (Prosil 28)	American Scientific, Sunnyvale, CA
Microtome (Reichert model 860)	Thomas Scientific, Swedesboro, NJ
Dermatome (Brown model 901)	Zimmer-USA, Warsaw, ID
Vapor trap absorbant (Tenax, GC, 60-80 mesh)	Alltech, Los Altos, CA
Flowmeters for monitoring airflow, (model No. 10)	Gilmont Instruments, Great Neck, NY
Fraction collector (model 10429)	Van Kel Industries, Edison, NJ
Drierite column	W. A. Hammond Drierite, Co., Xenia, OH
Tissue oxidizer (model OX-300)	Harvey Instruments, Hillsdale, NJ
Tissue oxidizer (model B-306)	Packard Instruments, Downers Grove, IL

REFERENCES

1. **Simpson, W. L. and Cramer, W.,** Fluorescence studies of carcinogens in skin. II. Mouse skin after single and multiple applications of 20-methylchloranthrene, *Cancer Res.,* 5, 449, 1945.
2. **Middleton, M. C.,** New approaches to problems of dermatotoxicity, in *Dermatotoxicity,* Gorrod, G. W., Ed., Taylor and Francis Ltd., London, 1981, chap. 19.
3. **Maibach, H. I. and Feldmann, R. J.,** Regional variations in percutaneous penetration in man, *Arch. Environ. Health,* 23, 208, 1971.
4. **Neter, J. and Wasserman, W.,** *Applied Linear Statistical Models,* Homewood, IL, 492, 1974.
5. **Feldmann, R. J. and Maibach, H. I.,** Absorption of some organic compounds through the skin in man, *J. Invest. Dermatol.,* 54, 399, 1970.
6. **Feldmann, R. J. and Maibach, H. I.,** Percutaneous penetration of steroids in man, *J. Invest. Dermatol.,* 52, 89, 1969.
7. **Bartek, M. J. and LaBudde, J. A.,** Percutaneous absorption in vitro, in *Animal Models in Dermatology,* Maibach, H. I., Ed., Churchill-Livingston, Edinburgh, 1975, 103.
8. **Hawkins, G. S. and Reifenrath, W. G.,** Development of an in vitro model for determining the fate of chemicals applied to skin, *Fund. Appl. Toxicol.,* 4, S133-S144, 1984.
9. **Reifenrath, W. G. and Hawkins, G. S.,** The weanling pig as an animal model for measuring percutaneous absorption, in *Swine in Biomedical Research,* Vol. 1, Tumbleson, M. E., Ed., Plenum Press, New York, 1986, 673.

10. **Hawkins, G. S. and Reifenrath, W. G.,** Influence of skin source, penetration cell fluid, and partition coefficients on in vitro skin penetration, *J. Pharm. Sci.,* 75, 378, 1986.
11. **Bronaugh, R. L. and Stewart, R. F.,** Methods for in vitro percutaneous absorption studies. III. Hydrophobic compounds, *J. Pharm. Sci.,* 73, 1255, 1984.
12. **Riley, R. T. and Kemppainen, B. W.,** Effect of serum-parathion interactions on cutaneous penetration of parathion in vitro, *Food Chem. Toxic.,* 23, 67, 1985.
13. **May, R. S. and DeClement, F. A.,** Development of a radiometric viability testing method for human and porcine skin, *Cryobiology,* 19, 302, 1982.
14. **Stoughton, R. B.,** Percutaneous absorption, influence of temperature and hydration, *Arch. Environ. Health,* 11, 551, 1965.
15. **Poulsen, B. J.,** Design of topical drug products: biopharmaceutics, in *Drug Design,* Vol. IV, Ariens, E. J., Ed., Academic Press, New York, 1973, chap. 5.
16. **Black, K. E. and Jederburg, W. W.,** Athymic nude mice and human skin grafting, in *Models in Dermatology,* Vol. 1, Maibach, H. I. and Lowe, Eds., Karger, Basel, 1985.

Chapter 6

IN VIVO METHODS FOR DETERMINING PERCUTANEOUS ABSORPTION AND METABOLISM OF XENOBIOTICS: INDIRECT METHODS*

Larry L. Hall and P. V. Shah

TABLE OF CONTENTS

*   This paper has been reviewed by the Health Effects Research Laboratory, U.S. Environmental Protection Agency, and approved for publication. Approval does not signify that the contents necessarily reflect the views and policies of the Agency nor does mention of trade names or commercial products constitute endorsement or recommendation for use.

I. INTRODUCTION

The extent and rate of percutaneous absorption is frequently of interest in the course of investigating the physiology, pharmacology, or toxicology of a substance. In many instances, the investigator does not wish nor have the time to become an expert in skin absorption but would like to know if the substance that is being investigated can be absorbed through the skin. The explicit purpose of this chapter is to describe the indirect approaches and techniques which have been reported in the literature for indicating or measuring percutaneous penetration in living animal models. Results produced by indirect penetration methods imply ingress of the chemical through some measurement(s) external to the transport processes through the skin. Thus inferences are made about the dermal penetration process through measurement of the penetrant in blood, tissues or excreta. Confirmation of the rates and species of percutaneous penetration using a direct method may be indicated. Only selected literature has been cited in this chapter. Comprehensive reviews on the issues and general methods of percutaneous penetration can be found in the dermal literature.[1-12] Two dosing approaches have been described for use in determining the percutaneous penetration of a chemical. One is the infinite dose situation where the concentration or amount of substance applied to the skin insures no significant depletion of the chemical on the donor site and results in zero-order or steady-state kinetics. The infinite dose situation has been used primarily in studying skin permeation *in vitro* but this condition may occur with poorly absorbed chemicals and some pharmaceutical preparations, although the maximal dermal dosage under therapeutic usage conditions is only approximately 2 to 5 mg/cm². The other dosing condition, the finite dose situation, has been the method used for most *in vivo* studies as it approximates the usual clinical situation. A thin film or layer containing approximately 4 μg/cm² is produced on the skin surface using a volatile solvent. As depletion of the donor source occurs, steady-state conditions do not exist and the kinetics of penetration are not zero-order. The finite dosing situation is the one commonly used in *in vivo* studies, and the discussion in this chapter will focus on this situation. The methods and approaches described in this chapter include, surface disappearance after skin application, determination of biological response after acute dermal exposure, measurement of penetrant in biological samples after dermal application, integumental penetration in nonmammalian species, and the measurement of first-pass metabolism after skin exposure.

II. SURFACE DISAPPEARANCE

Measurement of the loss of a chemical from the surface of the skin after dermal application has been used to estimate skin penetration. Tape stripping,[13,14] histologic,[15] autoradiographic,[16] and radionuclide surface counting techniques[17-21] have been reported. The tape stripping technique is predicated on the removal of the test substance from the surface of the skin by peeling off the horny layer with cellophane tape 20 to 30 times at the application site and measuring the material removed on the pieces of tape. The histologic method utilizes biopsies of the application site at various times with microscopic identification of the location of the penetrant. The autoradiographic approach locates the radioactive molecules in the skin biopsy through the number of developed silver grains in the photographic emulsion.[22,23] External radiation counting techniques are used to quantitate the amount of nuclide remaining on and in the skin. The external counting technique is the most frequently reported method for measuring surface disappearance. While in principal, these methods appear reasonable for determining skin absorption, uncertainties associated with each approach limit their usefulness for determining percutaneous penetration. Each method does, however, provide information about the movement of substances within the skin.

A. ANIMAL SPECIES

Studies with rats, guinea pigs, dogs, and humans have been reported. In principal, any species can be used with these methods.

B. CHEMICALS

Chemical purity should be greater than 95%. Be aware that the impurities limit the sensitivity of the methods and introduce a confounding variable into the study. The histologic technique requires chemicals with special physical or optical properties so the substance can be seen microscopically. Diffusible unbound chemicals cause special problems and uncertainties with the histologic and autoradiographic technique due to the requirements for fixation and embedding of the tissue. Radioactive tritium (^{3}H), carbon (^{14}C), phosphorus (^{32}P), and sulfur (^{35}S) which emit beta particles with energies (Emax) of 18.6, 156, 1710, and 167 keV, respectively, are frequently used for autoradiographic studies. Gamma emitting isotopes and the beta emitting nuclides except for the low energy tritium have been studied for skin penetration using external counting techniques. Tritium labeled compounds are most useful for the autoradiographic technique because of the higher resolution due to the shorter range of the lower energy beta particle.

C. SOLVENTS

As solvents can have a profound effect on dermal penetration, careful consideration should be given to their choice. Acetone and ethyl alcohol are frequently used because of their volatility, good solvency properties, and their use produces a thin layer of the chemical at the application site. Even water affects skin absorption as it hydrates the skin at the application site. Specific issues are discussed in other sections of this chapter.

D. SKIN SITE PREPARATION

Any hair should be removed by clipping 24 h before the study is performed. Site protection between measurements may be necessary to prevent loss of material.

E. RESTRAINT

Physical or chemical restraint may be necessary in order to make the measurements. External radionuclide counting requires precise, reproducible counting geometry.

F. ANESTHESIA

Local anesthesia with xylocaine is necessary for skin biopsies, but should not interfere with the measurement. Chemical restraint may be necessary for external counting in experimental animals. The short acting barbiturates can be useful for this procedure.

G. EQUIPMENT

Stripping of the skin can be accomplished with cellophane or Blenderm Surgical tape (3M Co., St. Paul, MN). Thin window and pancake Geiger-Muller (G-M) probes (Victoreen Nuclear Associates, Cleveland, OH), (Bicron, model PGM or EWGM, Newbury, OH), (Eberline, model 210, Santa Fe, NM) are used to measure beta emitting isotopes. Scalars (Eberline, model MS-2, SRM-100, Santa Fe, NM), (Bicron Labtech, Newbury, OH) will record cumulative counts from the G-M probes with efficiencies of 10 to 15% for ^{14}C labeled chemicals with ideal counting conditions. Some device is necessary to prevent contamination of the probe. Stainless steel wire screen (10/in.) has been used successfully for this purpose.[21] Scintillation detectors (NaI(Tl) Teledyne Isotopes, Westwood, NJ) are used to detect gamma photons. Collimators are necessary to restrict the area of measurement. Photographic emulsion NTB-2 (Eastman Kodak, Rochester, NY) is used for autoradiography with beta particle energies up to 200 keV. Tissue is mounted in OTC Compound (Lab-Tek Products, Naperville, IL) and cut with a cryostat (IEC 3398B, Fisher Scientific, Pittsburgh, PA).

H. DOSING

The test chemical should be applied uniformly over the application area. The volume of solvent used should be minimal but sufficient to cover the exposure area without running off. Water solutions or suspensions may require gentle spreading with an applicator. If a volatile solvent is used, apply small drops in order to cover the application site without run off of the solution. Dosing volumes of 10 to 40 μl/cm² are usually used for dermal application. The skin dosage should be determined by the method of detection.

I. SAMPLE PREPARATION

Biopsies or skin samples for morphological analysis are prepared by standard paraffin or plastic methods, and by frozen section. Standard radioactive sources or prepared standards (New England Nuclear, Boston, MA) are necessary to calibrate the counting instruments.

J. PROCEDURES

1. Tape Stripping Technique

At selected intervals after dermal application, the demarcated area is cleaned with mild soap and water, dried with cotton swabs, and tape stripped 20 to 30 times by placing a piece of the tape on the site and rubbed gently with an applicator and then pulling the tape off of the skin. Each section or sometimes two pieces of tape are assayed for the chemical by the appropriate method. The results produce a picture of the chemical distribution through the stratum corneum with time.

2. Histologic Studies

Skin site biopsies or skin samples, taken at various times, are processed by the appropriate fixation and embedding method for the chemical under study. Solubilization of the chemical under study must be prevented or minimized. Frozen sections of the skin will preserve the attachment of the stratum corneum to the viable epithelium. Sections of 5 to 10 μm can be cut. The sections are then processed for chemical identification with histochemical, antibodies, or other chromophores for microscopic visualization. Densitometry measurement can be used to quantitate the chemical.

3. Autoradiography

This is a very useful but demanding technique. Consult the references for specific details concerning this method.[22,23] Briefly, the skin samples or biopsies taken at various times after dosing are mounted in OTC media, frozen in liquid nitrogen and cut with the cryostat at 6 to 10 μm. The sections are mounted on Kodak NTB-2 emulsion coated slides, stored in light tight boxes for 1 to 8 weeks, and developed. The slides are then stained with hematoxylin and eosin or other stains and examined microscopically. All of the emulsion coating, mounting, and exposure are performed in the dark or with a safe light. This is a very powerful technique for chemicals which bind to cellular elements while translocation of diffusible substances is minimized.

4. External Counting Technique

The accuracy of this method is dependent on the reproducibility of the counting geometry. At selected times after dosing, the appropriate radiation detector (G-M or NaI crystal) is positioned over the application site of the restrained animal or human and radiation is measured. G-M counting efficiencies for ^{14}C-labeled compounds of 5% at the surface, 3% at the base of the stratum corneum, and 0.45% at the basal layer are reported.[21] Monitoring background between counts will identify contamination of the detector. Photon counting will usually yield higher efficiencies with little loss with depth, but will provide less precise localization to the skin as activity in the underlying tissues will also be measured. Although dependent on collimation and photon energy, companion counts, if possible, at a similar site away from the application area may be useful for correcting for other tissue, e.g., blood, muscle, radioactivity. One thousand counts has a 3% standard deviation. In general, adjust the counting time to achieve statistically significant data.

K. DATA ANALYSIS

Standard statistical (ANOVA, regression) analysis with appropriate transformations will aid in the interpretation of the data. Decay correction may be necessary for the short half-life radioactive isotopes.

1. Commentary

The tape stripping technique determines only the retention and loss in the stratum corneum and, as such, does not measure percutaneous penetration. It is most productively used as an adjunct to the excretion or blood measurement techniques.

The histological and autoradiographic procedures are most useful in providing information on penetration profiles and the routes of absorption through the skin. Translocation of diffusible substances during tissue processing introduces uncertainties in interpretation of the results.

The disappearance curves produced by external counting may not accurately reflect percutaneous penetration. This results from the decrease in counting efficiency with depth of penetration (beta emitters) and from the radioactivity in tissues located beneath the skin when gamma emitting isotopes are used. Substantivity or the binding of the chemical to keratin in the stratum corneum may be determined by surface counting. Like the tape stripping technique, surface counting is most useful when used as an adjunct to the excretion or blood measurement procedures.

III. DETERMINATION OF BIOLOGICAL RESPONSE AFTER ACUTE DERMAL ADMINISTRATION

A biological response, whether local or systemic, in the organism occurring after application of the test substance to the skin indicates that the substance has traversed the skin. The time course of the response after skin application is a function of the penetration kinetics and the time course of the biological response itself. An early strategy to assess the effect of formulation on skin absorption used strychnine induced convulsions which occurred following dermal exposure to strychnine formulated in several vehicles.[24] The bioassay approach exploits any biological response including both local and systemic effects associated with the chemical under study that can be quantified, e.g., lethality, narcosis, convulsions, motor activity, skin erythema, skin blanching, etc. Systemic responses after skin exposure when compared to the response following parenteral (or oral) exposure provide qualitative estimate of the rate and extent of dermal absorption since parenteral administration provides 100% absorption at zero time. The time course of the biological response can be characterized independent of the absorptive process. Using two or more periods of dermal exposure, e.g., 4 and 24 h, may provide information on the rate of skin absorption. This bioassay approach has shown utility in the assessment of the toxicity of pesticides.[25-29]

A. ANIMAL SPECIES
The species and strain of animal used for the percutaneous studies should be the same as the biological model that is being investigated. This will provide the most functional information and obviates any uncertainties associated with extrapolation of the results of the study. Rodents, usually rats or mice, guinea pigs, rabbits, cats, dogs, pigs, and nonhuman primates (rhesus monkeys) have been used. Rabbits may be required for acute dermal toxicity tests which are submitted to government agencies. Hairless mice, rats, and guinea pigs are available (Charles River, Kingston, NY, Harlan SD, Inc, Indianapolis, IN) and have been found useful.

B. CHEMICALS
The test substance should be of sufficient purity to prevent any ambiguity in the interpretation of the results. Instances exist in the literature where contaminants have been shown to possess significant biological activity and compromise the test results.[30,31] In general, chemical purity of 95% or greater is desirable.

C. SOLVENTS
The solvent should have minimal systemic toxicity itself at the dose used and not alter the barrier properties of the skin. Dermal absorption depends on the solubility of the substance in the vehicle and the partitioning between the solvent and the skin. Water or volatile solvents such as acetone or ethyl alcohol are used because of their volatility and excellent solvation properties. If solution of the chemical is not possible, and a paste or suspension is used, then control of the particle size is necessary as this factor can significantly effect dermal absorption. Solvents used for parenteral administration need to also be nontoxic at the doses used. Water, saline, propylene glycol solutions, or a mixture of nonionic surfactant (Emulphor EL-620, GAF Corp., New York, NY), ethanol and water (1:1:8), have been described. Parenteral administration of undiluted organic solvents is to be avoided. Even 10 μl of ethanol injected intraperitoneal will cause peritonitis and death. Solvent effects on dermal absorption can be large and therefore, careful consideration should be given to the choice of vehicle.

D. SKIN SITE PREPARATION
The hair on the dorsal application site should be gently clipped 24 h before treatment

with animal hair clippers using the size 40 blade (Oster Co., Milwaukee, WI). Approximately 10% of the body surface area, e.g., 10, 38, 62, and 220 cm^2 in adult mice, rats, guinea pigs, and rabbits, respectively, can be used for dermal application in biological response studies.[32] The site should be gently cleaned after clipping to remove dirt and sebaceous secretions using a mild soap solution (e.g., Liquid Ivory, Proctor and Gamble, Inc, Cincinnati, OH) and rinsed thoroughly with water.

E. RESTRAINT

Some form of restraining device which protects the application site is necessary to prevent oral intake and loss of the material from the dermal site of application. Elizabethan collars, jackets, sleeves, or other nonocclusive dressings have been used. Occlusive dressings, although not usually recommended can be considered as a way to maximize skin absorption through hydration of the horny layer and thereby enhance the sensitivity of the bioassay. Factors, e.g. chemical volatility, etc. may indicate occlusion of the skin site.

F. EQUIPMENT

Animal feeding needles (Popper and Sons, Inc., New Hyde Park, NY) are available for rodents and polyethylene catheters (No. 5 french infant nasogastric tube (American Hospital Supply Corp., McGaw, IL) in rabbits) can be used for gavage of larger animals, if the oral route is selected for comparison. The restraining and application site protection devices are usually improvised by the investigator.

G. DOSING

The volume of solvent used should be minimal but sufficient to cover the exposure area without running off. Water solutions or suspensions may require gentle spreading with an applicator. If a volatile solvent is used, apply small drops in order to cover the application site without run off of the solution. Dosing volumes of 10 to 40 $\mu l/cm^2$ are usually used for dermal application. Two to ten g/kg is the highest chemical dosage generally employed in acute toxicity studies.

H. PROCEDURES
1. Systemic Effects

Deichmann and LeBlanc[33] described a procedure for determining the approximate lethal (effective) dose with about six animals. Briefly, an estimate of the effective dosage range is determined from the substances chemical or physical properties or through the response of similar compounds. Graduated amounts, each 50% higher than the preceding one, are used to estimate the ED_{50} using one animal per dose. The lowest effective dosage is used as the estimator of the effective dose for 50% of the animals (ED_{50}). These authors reported agreement within $+33$ and -22 percent for the estimate of the ED_{50} determined with the standard protocol. If more precise information is needed, additional animals per treatment group can be added. The animals should be housed individually to prevent oral ingestion by cage mates and cannibalism should death occur. When appropriate, animals should be randomized for treatment group assignment. Exposure periods of up to 24 h are used and then the application site is washed with mild soap and water. Animals should be observed frequently during the first day and whenever signs are observed. Thereafter, observations twice per day at least 4 h apart for 14 d are usually sufficient for most compounds. Daily weights provide useful information. Record all signs of poisoning with time of appearance and duration.

A bioassay that has been quite useful in evaluating the dermal absorption and toxicity of pesticides, particularly the organic phosphorus and carbamate compounds, is the determination of blood cholinesterase activity.[33-36] This approach has also been very useful in assessing the exposure of agricultural field workers to these pesticides.[37] Pseudocholines-terase (EC 3.1.1.8 pChE) in serum or plasma and acetylcholinesterase (EC 3.1.1.7 AChE) in tissue and erythrocytes are determined after dermal exposure. Enzyme assay is by colorimetric[38] or electrometric[39] methods. Experimental duration may be up to three weeks if recovery is followed. Blood and tissues can be assayed in experimental animals using a serial sacrifice protocol. Longitudinal studies with multiple blood samples taken from the orbital sinus, caudal tail vein or tip of tail in rodents or marginal ear vein in rabbits have been reported.[40] Dermal absorption of other enzyme inhibitors could be assayed with this approach.

2. Epidermal Effects

Local pharmacological reactions at the application site have been used to study percu-taneous penetration and the factors which effect absorption in clinical subjects.[41] Shelley and Melton[42] used the wheal reaction to histamine to investigate the role of selected factors on skin absorption. The factors included vehicle, concentration, chemical form, regional variation, occlusion, and individual variation. Nachman and Esterly[43] used the blanching response to phenylephrine to evaluate the skin permeability in preterm human infants. Albery et al.[44] employed the erythema response following skin application of nicotinic acid esters to determine percutaneous absorption and transport in the dermis. Local microcirculatory blood flow measured by Laser Doppler Velocimetry has been used to study skin irritation, penetration enhancers, and the kinetics of skin penetration in man.[45-49] The steroid vaso-constrictor bioassay in humans has been used to compare the bioavailability and intrinsic activity of topical corticosteroid preparations.[50-52] Standardization has made this bioassay an important clinical procedure for biopharmaceutical assessments of steroids.[3,53] Investigators should consult the clinical literature for details of these methods.

I. SAMPLE PREPARATION

Use recommended procedures for the specific assay. For the cholinesterase assay, the blood sample is separated into plasma and red blood cell fractions.

J. DATA ANALYSIS

If extended treatment groups (i.e., more than one animal per dosage) are used, analysis by the methods described by Litchfield and Wilcoxon[54] and by Finney[55] will provide statistics such as ED_{50}, 95% confidence intervals, standard deviations, and slope of the dose response curve, which can be used to compare treatments. Analysis of the time to appearance of symptoms may follow the rate of dermal absorption. The method of Litchfield and Wilcoxon is a convenient graphical procedure and the Finney probit analysis can be calculated manually or by computer. Probit and analysis of variance (ANOVA) with contrasts procedures are found in the Statistical Analysis System.[56]

1. Commentary

The acute bioassay methods have been most frequently used to compare different routes of administration and may provide the easiest and most direct approach to ascertain skin penetration of the chemical of interest. However, a negative response does not rule out skin absorption as a level of skin absorption below that producing acute effects may be significant with repeated exposure. A quantitative test is needed to establish the rate and extent of dermal penetration. Only limited information is reported on the utility in dermal research or Laser Doppler Flowmetry and the steroid blanching response in experimental animals.

IV. MEASUREMENT OF PENETRANT IN BIOLOGICAL SAMPLES

A. MEASUREMENT OF BLOOD AND TISSUE LEVELS OF CHEMICALS FOLLOWING APPLICATION TO THE SKIN

Substances penetrating the epidermis diffuse to the capillary bed located approximately 200 μm beneath the skin surface and are carried away in the blood.[3] The virtual "sink" created by the large blood volume (7% of body weight) generally maintains the concentration gradient for diffusion after dermal application. Blood is the ideal medium for measuring disposition and metabolism in the body because of its homeostatic role in the organism. However, the usefulness of this medium has been limited by vascular access and analytical sensitivity. Advancements in analytical methodologies[57] and the use of radionuclides[58] have made it possible to quantitate the levels in blood found after dermal application. Improvements have also occurred in materials and methods for vascular access and collection of body fluids.[59-61] These changes, improved vascular access and increased analytical sensitivity, make it possible to perform studies even in the laboratory rat which has a blood volume of only approximately 20 ml (7% of body weight). Two approaches are used to determine dermal penetration by measurement in the blood. One is the serial sacrifice design,[62] used primarily with mice and rats. In this approach, groups of animals, usually 3 or 4 animals, are killed at various times after dosing and blood and tissues are taken for analysis. This protocol is the easiest to perform but the most difficult to analyze kinetically because of variability in the data introduced by intersubject variation. The longitudinal protocol, in which multiple blood measurements (and excreta as well) are made on the same animal is the most sensitive and accurate method for determination of distribution, excretion, metabolism,[63] and the rate and extent of dermal of absorption.[64-71] This approach is, however, technically more demanding, but the results can be worth the effort. If large animals, e.g., dogs, pigs, nonhuman primates, or cows, are used, vascular access is much simpler, larger blood samples can be taken, and loss of blood volume is not as severe a problem.

The concentration or amount of a substance in a tissue can under certain conditions be used to monitor dermal penetration. Poiger and Schlatter[72] used the percentage of the dose found in the liver at 24 h to determine the effects of route, i.e., both oral and dermal, and formulation on the absorption of tetrachlorodibenzodioxin (TCDD). Laug and Kunze[73] used the retention of lead in the kidney, a "biological magnifier", to show that this heavy metal is absorbed through the skin. The relationship of uptake and retention with time and dose is necessary with the use of this approach.

1. Animal Species

Small rodents, (mice, hamsters, rats), are usually used in serial sacrifice experiments. Rats, rabbits, cats, dogs, pigs, primates, and other large animals have been used for longitudinal studies.

2. Chemicals

Radiolabeled chemicals should have the radioactive nuclide label in a stable part of the molecule or that part of the molecule of interest. If only radiation detection is used, high purity material is needed. Radioactive impurities may compromise studies when low dermal absorption occurs. Measurement by radioactivity is sufficient to quantitate percutaneous absorption, but is limited for assessing metabolism and disposition. Sensitive analytical procedures permit more flexibility in chemical purity.

3. Solvents

The same considerations on the choice of solvent discussed in the biological response section apply. Careful thought should be given to the choice of solvent(s) if both parenteral and skin dosing are being used due to solvent effects not only on absorption but on other processes (distributive) as well. Solvents used for intravenous administration should be assayed for blood compatibility as hemolysis may alter distribution of the chemical.

4. Skin Site Preparation

Careful clipping is preferred as shaving and use of depilatories cause some degree of damage to the skin. Of the body surface area 1 to 3% is frequently used for the skin application area. If indwelling catheters are used, consideration should be given to the proximity of the sampling port to the skin application area.

5. Restraint

Restraint to prevent oral intake and loss of material from the application site on the skin cannot be over emphasized. Additional restraint may be necessary if indwelling catheters are used.

6. Anesthesia

Anesthesia in laboratory animals has recently been reviewed.[74] Light short-term anesthesia with medical grade ethyl ether in a closed anesthesia chamber provides useful restraint for clipping the hair from the backs of rodents and during application of radioactive chemicals. Induction and maintenance concentrations of 15 to 20 and 3 to 10 vol%, respectively, are suggested. If the concentration of ether is too high, breath holding occurs, induction will be prolonged and death may occur. This occurs most often when the chamber is first charged. Care should be taken to prevent contact with the skin. Ether should be used only in a well ventilated space (fume hood) and precautions should be taken to preclude peroxide formation during storage. Anesthesia for minor surgical procedures as implanting vascular catheters can be achieved with a number of agents. Ketamine/xylazine mixtures administered intraperitoneal provide good anesthesia and analgesia. Studies of the efficacy of dosing by the intraperitoneal route report injection into the intestinal space or into other internal organs in up to 20% of the attempts. Careful observation is necessary to compensate for this error.

Dilution of the anesthetic by one half is good practice as this allows more accurate measurement of the dosing volume, facilitates absorption, and reduces the possibility of overdose. Knowledge of the signs of anesthetic stages is important as the signs differ between species. In rabbits, loss of corneal reflex is a sign of dangerously deep anesthesia. Fasting the animal 8 to 12 h before treatment is a good practice. Since small animals lose body heat rapidly, some provision for maintaining body temperature is necessary. Postoperative intraperitoneal injection of saline (0.01% body weight) will enhance recovery.

7. Equipment

Medical grade Silastic tubing (Dow Corning Corp, Midland, MI) is a good material for indwelling catheters because this polymer causes less foreign body reaction in the tissues. Care is necessary with ligatures around silastic cannulas because of its softness. Indwelling catheters for larger animals are available (canine 18 ga., feline 20 ga., Monoject, St. Louis, MO). Restraining chairs (PLAS-LABS, Lansing, MI) provide effective control for nonhuman primates. Conditioning the monkeys reduces the stress associated with the chair restraint.

8. Dosing

Nonlinear dose absorption has been reported[62] and therefore the amount of chemical applied to the skin should be considered carefully. Excessive amounts should be avoided. A dosage of 4 μg/cm^2 has been used in a number of human and animal studies. The volume of solvent used to deliver the dose should be minimal but sufficient to cover the exposure area without running off. Water solutions or suspensions may require gently spreading with an applicator. If a volatile solvent is used, apply small drops with a microliter syringe (Hamilton, Reno, NV) in order to cover the application site evenly without runoff of the solution. Dosing volumes of 10 to 40 μl/cm^2 are commonly used for dermal application. Sufficient radioactivity should be used to detect low levels of poorly penetrating substances. In studies with rodents, 2 to 40 μCi of ^{14}C-radioactivity/kg body weight are commonly used. The amount of solvent for the parenteral dose should be minimal, nontoxic, and compatible with blood. In mice, vasculature access is difficult and intramuscular injections may be an alternative to the intravenous route. The intravenous dose of the chemical administered should be similar to the dermal dose so that any nonlinearities in disposition will not effect the results.

9. Procedures

Total blood loss from sampling should be less than 10% to prevent significant hemo-dynamic shifts which will introduce uncertainties in the data. A sampling procedure which accommodates the dead volume in the sampling system is necessary for the collection of valid blood samples. Consult the literature for specific information concerning the equipment and procedure for a particular animal model.[60] Briefly, the procedure for cannulating the jugular vein in the rat[59] or rabbit[61] is as follows. Anesthetize the animal to permit a 45 min procedure. Using standard aseptic procedures, expose the jugular vein and implant the sterile catheter, which is filled with heparinized saline, posterior to the bifurcation and secure with ligatures. The vascular assess port is then directed dorsally in the subcutaneous tunnel and exteriorized between the shoulder blades. Cannula care is necessary to maintain patency of the catheter and prevent infections. Recovery periods of 1 to 10 d are reported and vary with species and experimental duration. After recovery, the animal is dosed, placed in a metabolism cage, and the blood is periodically sampled. A preliminary study is useful for estimating the experimental duration. Dose three animals and collect excreta for 72 h. If 75% or more of the dose is excreted during the 3 day period, sample at 0.5, 1, 2, 4, 8, 24, 48, and 72 h. If between 50 and 75% is excreted in the 72-h period, add a 96- and 120-h samples. If less than 50% is excreted during the 3-d period, add a 10- and 20-d time point.

10. Sample Preparation

Measurement of the unbound fraction in plasma or serum is desirable. Ultrafiltration methods facilitate this determination (Centrifree, Amicon, Danvers, MA). Sample oxidation (Packard Inst. Co., Downers Grove, IL) is the most accurate and sensitive method for determining radioactive carbon and hydrogen in biological samples. Long liquid scintillation counting times, 60 min or more, may be necessary to quantitate the low levels found in blood.

11. Data Analysis

The extent of dermal absorption can be determined by using the area-under-the-curve (AUC) method using the parenteral [intravenous (i.v.) or intramuscular] and dermal plasma data by the following equation:

Dermal absorption (fraction) = AUC (dermal) $\times$ dose (i.v.) / AUC (i.v.) $\times$ dose (dermal)

The AUC can be measured by (1) the gravimetric integration method using the weight of the cut out plasma time curves, (2) using a planimeter, (3) or calculated using the trapezoid rule:

$$\text{area} = 1/2 \ (y_0 + y_1) \ (x_1 - x_0) + 1/2 \ (y_1 + y_2) \ (x_2 - x_1) + 1/2 \ (y_{n-1} + y_n) \ (x_n - x_{n-1})$$

where y_1 = blood concentration at ordered time points x_1. Only the cumulative dermal penetration is determined by the AUC method. If the rate of dermal absorption is of interest, another approach to the data analysis must be used. Deconvolution methods can be used to determine the rate of percutaneous absorption.[75,76] Briefly, the deconvolution problem consists of determining the dermal absorption rate with time knowing the impulse response, i.e., serum levels after an i.v. dose, and the system response, that is the serum levels after dermal exposure. Deconvolution procedures are described.[77] The rate of dermal absorption can also be determined by compartmental pharmacokinetic analysis[65,66,69] or with physiologically based pharmacokinetic models[62,78] using SAAM27 (Simulation, Analysis, and Modeling[79]) and other computer programs.

12. Commentary

This method is the most technically demanding, both in terms of the biological preparation and the analytical chemistry. The most frequently encountered difficulty is low blood levels after dermal application of slowly absorbed chemicals. Measurement of the chemical in blood after dermal application using the longitudinal protocol is the most accurate procedure to determine the rate and extent of percutaneous penetration.

B. MEASUREMENT OF EXCRETION AFTER DERMAL APPLICATION:

Measuring the appearance of a substance in the urine after skin application is the most widely used method for the determination of dermal absorption. This abbreviated deconvolution approach is especially practical in humans, and is commonly known as the "indirect method". The deconvolution problem consists of determining the dermal absorption rate with time knowing the impulse response, i.e., urine levels after an i.v. dose, and the system response, that is the urine levels after dermal exposure. This deconvolution approach along with the Mass Balance technique are perhaps the easiest methods for measuring dermal absorption in humans and laboratory animals. Details of this method can be found in a number of articles.[5,58,80-88] In this approach, a test chemical is applied to the desired area of the skin and the appearance of radioactivity in the urine is measured. As excretion in urine may not be the only route of elimination, a equivalent absorbed dose is given parenterally and urine is collected until excretion is complete. Similar dosages are necessary in order to maintain linear kinetics. The extent of dermal absorption is then calculated by correcting the apparent skin penetration for incomplete urinary excretion. If the rate of excretion after parenteral administration is much greater than the rate of skin absorption, then the rate at which the chemical appears in the urine after application to the skin will approximate the rate of dermal absorption.[80]

1. Animal Species

The choice of the animal species to use for measuring dermal absorption should be dictated by the anticipated use of the information. This indirect method has been employed in most experimental laboratory animals as well as in humans.

2. Chemicals

The test substance should be of sufficient purity to prevent any ambiguity in the interpretation of results. In general, chemical purity of 95% or greater is desirable to preclude any interactive effects between ingredients. Use of radiolabeled material simplifies the analysis (quantitation) of urine for the penetrant. A radiochemical purity of 98% or greater is desirable. Periodical purity check is required to detect radiolysis or other decomposition of the chemical. The location of radioactive tag in the molecule should be at a stable position that will permit tracking that portion of the molecule of interest to the investigator.

3. Solvents

Careful consideration should be given to the choice of solvent used in the study. Acetone and ethyl alcohol are frequently used and have become somewhat the de facto standards. This use can be attributed to the high solvency and the rapid volatility of these vehicles. While it is recognized that this method of application is unnatural, the use of these two solvents provides a standard method of application which tends to normalize the effect of the solvent in dermal penetration studies and provides a basis for comparison of other vehicles of application. For parenteral (i.v.) administration, solvents such as normal saline, propylene glycol solutions and a mixture of nonionic surfactant, ethanol and water (1:1:8) have been described. Biocompatibility of the parenteral solvent system should be ascertained.

4. Skin Site Preparation

For studies with human volunteers, there is no special need for preparing skin for dermal absorption studies unless the subject is hirsute or a site other than the ventral forearm is used. Washing the skin site several hours before dosing with a mild soap and water has been suggested. Skin site preparation for laboratory animals includes removal of hair by clipping and washing. Briefly, the hair on the dorsal application site (mice, rat, hamster, guinea pigs) should be gently clipped 24 h before treatment with animal hair clippers using size 40 blade (Oster Co., Milwaukee, WI). Use of light ethyl ether anesthesia to prevent animal movement is recommended. The site should be gently cleaned after clipping to remove dirt and sebaceous secretions using a mild soap solution.

5. Restraint

Some form of device which protects the application site is necessary to prevent oral intake and loss of material due to rubbing. For human studies, non occlusive semirigid polypropylene Hilltop Chambers (Hilltop Research Inc., Cincinnati, OH) attached with hypoallergenic adhesive tape have been used successfully.[89] Perforated plastic blisters from Cathavex filters (Millipore Corp, Bedford, MA) and disposable beakers (Fisher Scientific, Pittsburgh, PA) affixed with cyanoacrylate adhesive has been effectively used in rats.[90] Teflon rings or other suitable devices, are commercially available (Crown Glass Company, Sommerville, NJ). Bartek et al.[83] have described a method to protect the skin site in animals. Many other appliances fabricated from commonly available materials in the laboratory have been successfully employed. Tygon or latex rubber tubing collars have been used[90,91] to protect the blisters from assault by the animal.

6. Anesthesia

Light ethyl ether anesthesia is helpful during skin site preparation and also during application of a test compound in rodents. Care should be exercised to minimize contact and duration of anesthesia. Ketamine/xylazine mixture administered intraperitoneally also provide good anesthesia and analgesia.

7. Equipment

Metabolism cages (Nalge Co., Rochester, NY) for collection of excreta from rodents during the treatment period are used. Metabolism cages for other species are available (Allentown Caging Equipment Co., Allentown, NJ).

8. Dosing

The amount of chemical applied to the skin is dependent upon the purpose of the study. Excessive amounts should be avoided. A dosage of 4 $\mu g/cm^2$ has been used in a number of human and animal studies. This is the amount of active ingredient that would be deposited from a 0.25% cream used in a clinical setting. The volume of solvent used to deliver the dose should be minimal but sufficient to cover the exposure area without running. Water solutions or suspensions may require gently spreading with an applicator (nonabsorbing type). Heat sealed polyethylene tubing has been successfully used to spread water solutions evenly on rat skin.[90] If a volatile solvent is used, apply small drops with a microliter syringe (Hamilton, Reno, NV) in order to cover the application site evenly without runoff of the solution. Dosing volumes of 10 to 40 μl/cm^2 are commonly used for dermal application. Sufficient radioactivity should be used to detect low levels of poorly penetrating substances. In studies with rodents, 2 to 40 μCi of ^{14}C-radioactivity/kg body weight are commonly used. The amount of solvent for the parenteral dose should be minimal, nontoxic, and compatible with blood. In mice, vasculature access is difficult and intramuscular injections may be an alternative to the intravenous route. The dose of the chemical administered should be small to minimized any inherent toxicity. One microcurie radioactivity have been often used in humans for intravenous and dermal administration.[80,81] Hydration is important in human studies so that adequate urine is produced. Informed consent and institutional approval are required for human studies.

9. Procedure

In humans, the ventral forearm skin is frequently used for dermal application. Dosing solution is delivered to the marked site drop by drop with a Hamilton syringe or other positive displacement devices. Use of air displacement type pipettes to deliver the dose in a volatile solvent is not advisable because suction may cause rapid evaporation of the solvent. Following the application of the chemical, a perforated Hilltop Chamber (Hilltop Research Inc., Cincinnati, OH) or other protecting device is secured over the treated area. Hypoallergenic adhesive tape has been used to secure the chamber in human studies.[89] If a test chemical is highly volatile, then an occlusive device (chamber without holes) may be indicated. The excretory products (usually urine) are collected until excretion of the penetrant is complete. Treatment duration is dependent upon the penetration and excretion characteristic of the test material. Treatment duration of 5 to 7 d is frequently satisfactory. Longer duration may be necessary for the very lipophilic and slowly metabolizing chemicals like the polychlorinated biphenyls. In a second part of the study, the same subjects (following achievement of background urinary levels) are given an parenteral (i.v.) dose of the test compound and urine is collected daily until excretion is complete. The first day is divided into three 4-h periods followed by a 12-h period.

This technique has also been used in laboratory animals. Light ether anesthesia during skin site preparation and application of a test chemical is suggested in order to prevent skin abrasion during hair clipping and allow accurate dosing. Usually the mid-dorsal region of mice, rats, and guinea pigs is preferred for skin applications. The size of the treatment area is approximately 3% of the body surface area. Fur on the back of the animal should be clipped at least 24 h prior to treatment (see "skin site preparation" of this section). The treatment area is marked and the solvent containing the appropriate dose is delivered drop by drop with a Hamilton syringe or other positive displacement device. Care should be taken to prevent runoff during application. Perforated plastic blister (for example Cathavex single-use filters, Millipore Corp., Bedford, MA) or other appropriate device is carefully glued over a treated area with cyanoacrylate adhesive. Dislocation of the blister by the animal is prevented by placement of a collar of rubber tubing behind the forelegs as described by Bartek et al.[83] Details of animal preparation can be found in number of articles.[83,90,91]

It is desirable to house treated animals in metabolism cages (Nalge Co., Rochester, NY, and Crown Glass Co., Sommerville, NJ) for collection of excreta during the treatment period. For intravenous administration, the tail vein is recommended in rats and the marginal ear vein is commonly used in rabbits. The dose in an appropriate biocompatible solvent is infused in the vein and the treated animals are placed in metabolic cages to facilitate the collection of excreta. Excreta are collected at 6, 12, and 24 h the first day and then daily until excretion is complete.

10. Sample Preparations

Radioactive urine samples (0.1 to 1 ml) can be assayed directly using 10 to 15 ml of scintillation fluid (Insta-Gel, Packard Instrument Co., Downers Grove, IL.). Quench correction is achieved with either the known addition method (adding an internal standard eg. (^{14}C-toluene or *n*-hexadecane), and/or by the automatic external standardization procedure. Fecal samples are air dried, weighed, and ground (mini coffee grinder, Moulinex Products Inc., Virginia Beach, Va.) to obtain a homogeneous specimen. Aliquots of the fecal samples (100 to 250 mg) are combusted in a biological oxidizer (e.g., Packard Tri-Carb Sample Oxidizer, Packard Inst. Co. Downers Grove, IL., Harvey Biological Oxidizer, Harvey Instrument Corporation, Hillsdale, NJ). The decays per minute (DPM) should be determined on all radioactive samples. A routine quality assurance program is recommended.

11. Data Analysis

Dermal absorption is calculated with correction for incomplete excretion by the following formulae:

% Dermal absorption = fraction of dose excreted (dermal) × 100/fraction of dose excreted (i.v.)

Deconvolution analysis[75] or the use of pharmacokinetic models[92] may permit calculation of the rate of dermal absorption. The choice of the method of analysis is dependent upon the purpose and design of the experiment. Statistical analysis is a valuable aid in interpreting the results of an experiment.

12. Commentary

The "indirect method" as name implies, is an circuitous way of determining percutaneous absorption of a chemical. This method has permitted experiments with human subjects where sampling of tissues is not possible. It is a relatively simple method which is rapid and perhaps, less labor intensive than other methods. With this method an assumption is made that the elimination profiles after topical and parenteral administration are identical. Only total radioactivity is determined, which does not separate parent chemical and metabolites. Little information is obtained on the disposition or metabolism of a chemical following dermal absorption. It will not provide information on true percutaneous absorption rate of chemical unless excretion is much faster than dermal absorption. Uncertainties in the estimate of the extent of percutaneous absorption arise because of the use of the nominal rather than actual applied dose and the necessity to correct for incomplete urinary excretion.

C. "MASS BALANCE" TECHNIQUES

In this method, total accountability (recovery) of the chemical is accomplished. All tissues including carcass, excreta, application site and protective devices, and all volatiles are captured and assayed. Accounting for all of the delivered dose reduces the experimental variability caused by errors in conducting the experiments. This method has been frequently used in small laboratory animals, e.g., mice, rats and guinea pigs, to determine the rate and extent of percutaneous absorption, and disposition of the chemical following dermal application.[90,93-100] Recently, the mass balance approach has been applied to a limited extent in human experimentation with chemicals whose elimination occurs entirely in the urine.[89] This mass balance method and the excretion analysis method are the most commonly used techniques for determining percutaneous absorption in humans and experimental animals.

1. Animal Species

The choice of the animal species to use for measuring dermal absorption should be dictated by the anticipated use of the information. This "mass balance" technique has been employed in most experimental laboratory animals and to a limited extent in humans.

2. Chemicals

The test substance should be of sufficient purity to prevent any ambiguity in the interpretation of results. In general, chemical purity of 95% or greater is desirable. Use of radiolabeled material simplifies the processing of biological samples for the quantitation of the penetrant. A radiochemical purity of 98% or greater is desirable. Periodic purity checks are required to detect radiolysis and other decomposition of the chemical. The location of radioactive label in the molecule should be at a stable position that will permit tracking that portion of the molecule of interest to the investigator.

3. Solvents

Desirable properties of the solvent for percutaneous studies have been discussed. Careful consideration should be given to selecting the vehicle as solvents can have profound effects on percutaneous penetration. Ideally, the solvent should have no systemic toxicity, high solvency, rapid volatility, and causes minimal alterations in the skin. Acetone and ethanol have been extensively used for percutaneous absorption studies in both humans and experimental animals.

4. Skin Site Preparation

Skin site preparation for laboratory animals includes removal of hair by clipping and washing. Briefly, the hair on the dorsal application site (mice, rat, hamster, guinea pigs) should be gently clipped 24 h before treatment with animal hair clippers using the size 40 blade (Oster Co., Milwaukee, WI). Bronaugh et al.[101] have shown that the dorsal skin on the male rat is thicker than abdominal skin while no difference was noted between dorsal and ventral skin thickness in the female rat. Use of light ethyl ether anesthesia to prevent animal movement is recommended. The site should be gently cleaned after clipping to remove dirt and sebaceous secretions using a mild soap solution.

5. Restraint

Some form of device which protects the application site is necessary to prevent oral intake and loss of material due to rubbing. Perforated plastic blisters from Cathavex filters (Millipore Corp, Bedford, MA) and disposable beakers (Fisher Scientific, Pittsburg, PA) affixed with cyanoacrylate adhesive has been effectively used in rats.[90] Teflon rings or other suitable devices, are commercially available (Crown Glass Company, Sommerville, NJ). Bartek et al.,[83] have described a method to protect the skin site in animals. Many other appliances fabricated from commonly available materials in the laboratory have been successfully employed. Tygon or latex rubber tubing collars have been used[90,91] to protect the blisters from assault by the animal. In humans, the ventral forearm skin is frequently used for dermal application. Following the application of the chemical, a perforated Hilltop Chamber (Hilltop Research Inc., Cincinnati, OH) or other protecting device is secured over the treated area. Hypoallergenic adhesive tape has been used to secure the chamber in human studies.[89]

6. Anesthesia

In experimental animals, light ethyl ether anesthesia is helpful during skin site preparation and also during application of a test compound in rodents. Ketamine/xylazine mixture administered intraperitoneally also provide good anesthesia and analgesia.

7. Equipment

Metabolism cages for collection of excreta from rodents during the treatment period are used. Metabolism cages for other species are available.

8. Dosing

Criteria for selecting appropriate dose have been discussed. Briefly, the amount of chemical applied to the skin is dependent upon the purpose of the study. Excessive amounts should be avoided. A dosage of 4 μg/cm^2 has been used in a number of human and animal studies. This is the amount of active ingredient that would be deposited from a 0.25% cream used in a clinical setting. The volume of solvent used to deliver the dose should be minimal but sufficient to cover the exposure area without runoff. Water solutions or suspensions may require gently spreading with an applicator (nonabsorbing type). Heat sealed polyethylene tubing has been successfully used to spread water solutions evenly on rat skin.[90] If a volatile solvent is used, apply small drops with a positive displacement microliter syringe in order to cover the application site evenly without runoff of the solution. Dosing volumes of 10 to 40 μl/cm^2 are commonly used for dermal application. Sufficient radioactivity should be used to detect low levels of poorly penetrating substances. In studies with rodents, 2 to 40 μCi of ^{14}C-radioactivity/kg body weight are commonly used. One microcurie of ^{14}C-labeled chemical has been used in human studies.[80,81]

9. Procedure

In this method with small laboratory animals, the serial sacrifice protocol is usually followed, i.e., group of animals are killed at desired time points following skin application. Three to four animals per time point are used with at least 5 time points. Excretory products are collected daily for analysis. The skin application site is often washed with soap and water at the end of the experiment and/or at 24 h post-application and the wash analyzed for its content. At the time of sacrifice, the protective device is removed and the skin application site is excised and both are analyzed for chemical content. Plasma, tissues or organs and the remains of the body (carcass) are taken for analysis and the total radioactive recovery for each animal is determined.

Recently, the excretion method for skin absorption measurement in human have been modified to achieve mass balance.[89] Experiments in human subjects are conducted as described previously in the measurement of excretion after dermal application section of this chapter. At 24 h after dermal application, the Hilltop chamber is removed and treated skin site is washed using cotton balls with soap and water (Ivory liquid soap, diluted 1:1 with water) followed by a rinse. This process is repeated and procedure is completed with two final rinse with water. Radioactivity in washes and sequestered with Hilltop chamber is determined. A second Hilltop chamber is placed over the application site following washing for the remaining period. At the end of the experiment, the Hilltop chamber is removed and analyzed for its content. The stratum corneum of the application site is stripped using cellophane tape and the tape strips assayed for chemical. The disposition of radioactivity into various fractions is calculated based on either the nominal or the recovered dose and percent absorbed dose is calculated as described previously.[89]

10. Sample Preparations

Radioactive urine samples (0.1 to 1 ml) can be counted directly for radioactivity using 10 to 15 ml of scintillation fluid (Insta-Gel). Fecal samples are usually air dried, weighed, and ground (hand ground by using pastel and mortar or ground in a minicoffee grinder to obtain a homogenous specimen. Aliquots of the fecal samples (100 to 250 mg) are combusted in a biological oxidizer (e.g., Packard Tri-Carb Sample Oxidizer, Harvey Biological Oxidizer,). Radioactivity in the protecting device (blisters, disposable plastic beakers, Hilltop chamber, etc.) can be determined by adding scintillation fluid and assaying in a scintillation counter. Aliquots of washes can also be counted for radioactivity by adding appropriate scintillation counting fluid. Treated skin can be cut into 8 to 10 pieces (less than 100 mg) and oxidized as mentioned above. Carcasses can be cut into 10 to 12 pieces and plunged into liquid nitrogen and blended in a blender (Waring Corp.) with a hole in the lid to reduce the pressure inside the chamber. The resultant hygroscopic powder is resuspended in water and blended to make a slurry. Aliquots of the homogenates can be combusted as mentioned above, and counted for radioactivity. Tissue samples should be minced and weighed. Aliquots from tissues (100 to 250 mg, depending upon radioactivity) are combusted, and counted for radioactivity. All radioactive counts should be converted to dpm by correcting for counting efficiency of the instrument and for quenching. A daily quality control procedure is recommended.

11. Data Analysis

The total DPM recovered is determined by summing the radioactivity (DPM) detected into each fraction. Percent dose recovered (total radioactive recovery) can be calculated based on the theoretical dose applied to the skin. Fraction of the dose recovered in tissues and excreta can be calculated based on either the theoretical dose applied to the skin or the total recovered dose. Further choice of the method of statistical analysis is dependent upon the purpose of the experiment, and its subsequent extrapolation. Analysis of variance models using tissue content and concentration and time of sacrifice will aid in estimating dermal absorption and identifying differences in tissue retention and kinetics. Pharmacokinetic modeling of the results will require a thorough analysis of the data and use of commercial software packages.[79]

12. Commentary

Total radioactive recovery (mass balance) accurately defines the delivered dose and recovered dose. It provides a direct check on experimentation. It enables the investigator to determine percutaneous absorption rates and extent as well as the disposition of a chemical following dermal absorption. Metabolism can be quantified. It does not assume the same excretion profile between different routes and determines absorption directly. It is a rather labor intensive method due to tissue processing and number of samples. However, the labor can be reduced if interest is only in the rate and extent of percutaneous absorption. The protocol calls for serial sacrifice which requires large number of animals. With the addition of chemical speciation, this is the most robust method to determine the disposition of a chemical after dermal application.

V. INTEGUMENTAL PENETRATION IN NONMAMMALIAN SPECIES

Knowledge of absorption following contact is important in pest control practices as well as for adverse action in nontarget species coexisting in the ecosystem. A good cuticle penetrant property of insecticide is desirable in insect control due to large body surface area to mass ratios of insects, complexity of appendages and easy accessibility of nervous system. Other nontarget species such as amphibia, reptiles, birds, and aquatic and terrestrial animals become the victims of man-made environmental pollution. Knowledge of absorption is desirable to protect the nontarget species from detrimental effects of pollutants. Absorption studies in insects,[102,103] frogs,[102,104] and birds[102,105-108] have been conducted. Included here is the narrative description of methods utilized in conducting integument absorption studies in nonmammalian species.

A. COCKROACH

Anesthetize roaches with CO_2 and apply the test chemical in a volatile solvent (usually acetone) to the pronotum. Total volume of solvent should not exceed 10 μl. Treated roaches can be held individually in pint glass jars covered with screen. Food and water may be withheld during the holding period. Hemolymph (blood) can be drawn at the desired time point by cutting antennae off, sealing the mouth part with hot paraffin, and subject the insect body to low centrifugal force. Cuticle at the application site is removed, rinsed with acetone and assayed for radioactivity. Other tissues can also be analyzed to determine the distribution. The holding jars can be rinsed with acetone to determine radioactivity in excretory products and any possible rub off.

B. HORNWORM

Anesthetize hornworms with CO_2 and apply the test chemical in a volatile solvent (usually acetone) to the dorsum of the first thoracic segment. Total volume of the solvent should not exceed 5 μl. The treated larvae can be held individually in pint glass jars covered with screen. Food may be withheld during the holding period. At desired time points, hemolymph (blood) can be obtained by removing the "caudal horn" (located at the posterior end) and collecting the fluid in a scintillation vial. Application site can be removed and analyzed for the content. The remainder of the body (carcass) can be homogenized, and assayed for the content. Fecal samples can be oxidized and counted for radioactivity.

C. JAPANESE QUAIL

The featherless area under the wing arm pit can be used for dermal application. An appropriate dose, dissolved in a volatile solvent, is delivered slowly with a Hamilton syringe. It is desirable not to exceed 50 μl total volume of solvent. The treated birds can be housed in rat metabolism cages containing filter paper to absorb (collect) excreta. At the desired time point, blood and other tissues can be harvested for radioactivity determination. The skin at the application site can be removed, rinsed with acetone and assayed for its content. Abou-Donia et al.[105,108] have used comb of hen for topical application.

D. FROGS

The test chemical in a small volume of solvent (10 μl) can be applied on the back immediately behind the head of grass frogs. Care should be taken to prevent running off of the dosing solution on the slimy surface. Treated frogs can be held in glass jars with a screen lid during the experiment. Use of ether anesthesia during treatment is recommended. At the end of the experiment, blood can be drawn by cardiac puncture. The application site as well as other tissues can be removed for radioactivity determination.

1. Sample Preparation

All samples can be processed for radioactivity determination as described under "mass balance" technique.

VI. METABOLISM

Although the skin is the second largest organ in the body, its metabolic capabilities are not well characterized. This is due, in part to (1) the difficulties in preparing the heterogenous tissue for metabolic examination, (2) the lack of well-defined afferent and efferent blood vessels which facilitate biochemical studies, and (3) the belief that the barrier function resides primarily in the dead horny layer. The capability for metabolism of xenobiotics by the skin has recently been reviewed.[109,110] Oxidative,[111-113] reductive,[114] hydrolytic,[67,112] and conjugation[112] biotransformation reactions have been identified in skin.

The determination of the metabolism of a chemical in transit across the skin does not have a unique *in vivo* solution, but rather is arrived at by difference, either in chemical species, time, or most often a combination of both events. The most unambiguous approach is to compare the dermal disposition and chemical speciation after dermal application with that after intravenous administration using radiolabeled penetrants. Wester et al.[63] applied this pharmacokinetic approach to determine the skin first-pass metabolism of nitroglycerin in the rhesus monkey. The chemical methods for detection of penetrants has been discussed in detail in separate chapters of this volume. In this section an approach for conducting dermal absorption-metabolism studies is presented. Knowledge of first pass metabolism is important in order to evaluate whether the parent chemical or its metabolite(s) are crossing the epidermis of the skin and entering the blood stream and role of the other layers of the skin in penetration. This information along with quantitative disposition and metabolism into various organs can be useful in developing pharmacokinetic models, and may aid in extrapolation of benefits and risk to humans.

Percutaneous absorption studies are conducted using appropriate method(s) for the animal model chosen to investigate first-pass metabolism in skin. Blood plasma, excreta, and tissues are collected at desired time points, and subsequently analyzed for identification and quantitation of the parent, any biologically active metabolites, and other significant metabolites. The following general outline may be adapted to achieve the objectives of the study.

A. Animal Species

Small rodents, (mice, hamsters, rats), are usually used in serial sacrifice experiments. Rats, rabbits, cats, dogs, pigs, primates, other large animals and humans have been used for longitudinal studies.

B. CHEMICALS

Radioactive labeled chemicals with chemical and radioactive purities of greater than 98% facilitate the determination of first-pass metabolism and disposition and minimize uncertainties in interpretation of the results. High purity unlabeled chemical simplifies the analytical problem and prevents any interactions with other chemicals which may effect dermal penetration.

C. SOLVENTS

Intravenous vehicles should be biocompatible with the blood to prevent artifacts in disposition kinetics. Vehicles for dermal application include acetone and ethyl alcohol. Pharmaceutical formulations (creams and ointments) have been used.

D. SKIN SITE PREPARATION

For studies with human volunteers, procedures described earlier using the ventral forearm are adequate unless a larger application area is necessary. Washing the skin site several hours before dosing with a mild soap and water has been suggested. Skin site preparation for laboratory animals includes removal of hair by clipping and washing. Gently clip hair 24 h before treatment with animal hair clippers using size 40 blade (Oster Co.). Use of light ethyl ether anesthesia to prevent animal movement is recommended. The site should be gently cleaned after clipping to remove dirt and sebaceous secretions using mild soap solution.

E. RESTRAINT

Some apparatus which protects the application site is necessary to prevent oral intake and loss of material due to rubbing. For human studies, nonocclusive semirigid polypropylene Hilltop Chambers (Hilltop Research Inc.) attached with hypoallergenic adhesive tape have been used successfully.[89] Perforated plastic blisters from Cathavex filters (Millipore Corp,) and disposable beakers (Fisher Scientific,) affixed with cyanoacrylate adhesive has been effectively used in rats.[90] Bartek et al.,[83] have described a method to protect the skin site in animals. Many other devices fabricated from commonly available materials in the laboratory have been successfully employed. Tygon or latex rubber tubing collars have been used[90,91] to protect the blisters from assault by the animal.

F. ANESTHESIA

Light ethyl ether anesthesia is helpful during skin site preparation and also during application of a test compound in rodents. Care should be exercised to minimize contact and duration of anesthesia. Ketamine/xylazine mixture administered intraperitoneal also provide good anesthesia and analgesia.

G. EQUIPMENT

Metabolism cages are necessary to collect and separate excreta and volatile materials. Collection vials and other labware should be evaluated for compatibility with chemical assay.

H. DOSING

Intravenous administration is used to characterize disposition after entrance in the blood stream. Do not use intraperitoneal administration as compounds given by this route are absorbed primarily by the hepatic portal circulation.[115] The dose applied to the skin is dependent upon the purpose of the study and nonlinearities in dose absorption should be considered. A dosage of 4 μg/cm^2 has been used in a number of human and animal studies. The volume of solvent used to deliver the dose should be minimal to prevent runoff. Dosing volumes of 10 to 40 μl/cm^2 are commonly used for dermal application. Sufficient radioactivity should be used to detect levels of the substance in plasma and tissues. A pilot study is encouraged to verify detectability in the tissue. In studies with rodents, 2 to 40 μci of ^{14}C radioactivity/kg body weight are used. One microcurie of ^{14}C labeled chemical has been used in human studies[80,81] using the urinary excretion method. Higher doses have been used in clinical studies[116,117] and must be approved by the Human Studies Committee(s). The amount of radionuclide used is controlled by the tissue radiation dose incurred during the experiment.

I. PROCEDURE

In this method with small laboratory animals, the serial sacrifice protocol is usually followed with 3 to 4 animals per group and at least 5 sacrifice times. Urine is collected at 4 to 6 h intervals for the first day. Excretory products are then collected daily for the duration of the study. The skin application site is often washed with soap and water at the end of the experiment and/or at 24 h post-application and the wash analyzed for its content. At the time of sacrifice, the protective device is removed and the skin application site is excised and both are analyzed for chemical content. Plasma, tissues or organs and the remains of the body (carcass) are taken for analysis.

Experiments in human subjects are conducted as described previously. At 24 h after dermal application, the Hilltop chamber is removed and treated skin site is washed. Radioactivity in washes and sequestered with the Hilltop chamber is determined. A second Hilltop chamber is placed over the application site following washing for the remaining period. At the end of the experiment, the Hilltop chamber is removed and analyzed for its content. The stratum corneum of the application site is stripped using cellophane tape and the tape strips assayed for chemical. Blood samples are taken frequently using an indwelling catheter until the plasma disposition is complete. Urine is collected quantitatively at 4, 8, 12, and 24 h the first day and then every 12 h for the duration of the study. Some procedure (e.g., freezing), during sample collection may be necessary to prevent decomposition of metabolites. Urine is sterile when produced, but is contaminated upon voiding.

J. SAMPLE PREPARATION
Enzyme or acid hydrolysis may be necessary if conjugates are present.

1. Analysis of Treated Skin
Treated skin from experimental animals can be excised at desired time points, and pooled if necessary. The skin is rinsed with acetone or other appropriate organic solvent to dislodged the compound on the skin surface. The skin rinse is concentrated by evaporation under nitrogen and/or freeze drying process and dissolved in small volume of appropriate solvent for further identification and quantitation of the test compound in this fraction. Then the treated skin is cut into small pieces and homogenized in an Omni Homogenizer or a hand (glass) homogenizer by adding distilled water (1:3). The skin homogenate is further extracted with organic solvent or a mixture of solvents (2 to 3 times) followed by distilled water (2 to 3 times) to remove both polar and non-polar fraction. The choice of solvent(s) is determined by solubility of the compound. Organic solvents such as heptane, ethyl acetate, ether, chloroform, benzene, and methyl alcohol can be used as an extracting solvents. A mixture of chloroform, acetone, methanol, water (8:2:2:1) has been shown to be a useful mixture for skin extraction.[118,119] These extracts can be concentrated by evaporation and/or freeze drying process and resuspended in a small volume of appropriate solvent for further identification and quantitation of the test compound in these fractions. Precolumn clean up prior to analytical identifications may be required. The residue of the skin is then analyzed to determine the bound (unextracted) quantity.

2. Urine Samples
Urine samples can be pooled from the same treatment group if necessary, hydrolyzed, and centrifuged at low speed (about 10,000 rpm for about 15 min) to precipitate solids, and concentrated by evaporation and/or freeze drying. Centrifuged urine samples can be directly subjected to analytical method(s) for identification and quantitation if sufficient quantity of the test chemical has been voided. Some analytical methods may require extraction of urine samples with polar and nonpolar solvents.

3. Feces Analysis
Feces samples at desired time point can be pooled from the same treatment group and air dried if volatiles are not present. Grind air dried fecal samples into fine powder by using mortar and pestle or using mini coffee grinder (Moulinex Inc.). Hydrolyze samples as indicated. Feces samples can be extracted with organic solvent or a mixture of solvents (2 to 3 times) followed by distilled water (2 to 3 times) to separate both polar and nonpolar fraction. A mixture of chloroform, acetone, methanol, water (8:2:2:1) have been used for extraction of feces.[118,119] These extracts are concentrated and resuspended in small volume of appropriate solvent for further identification and quantitation of the test compound in these fractions. Residue of the feces samples may be analyzed to determine the bound (unextracted) quantity. The radionuclide can be used to follow the extraction and analytical recovery of the sample.

4. Analysis of Biological Tissue
Blood, liver, fat, kidneys and other biological tissues may be extracted with appropriate solvents depending upon the test chemical. Plasma or serum is generally preferred over a whole blood and should be assayed for bound fraction using dialysis or ultrafiltration methods (Amicon, Danvers, MA). Tissue samples may be pooled from the same treatment group, if necessary. The pooled tissues are cut into small pieces and homogenized in an Omni Homogenizer or a hand (glass) homogenizer with distilled water (1:3). The tissue homogenate is extracted with organic solvent or a mixture of solvents (2 to 3 times) followed by distilled

water (2 to 3 times) to separate both polar and nonpolar fraction. Organic solvents used are heptane, benzene, ethylene dichloride or isobutanol can be used as an extracting solvents. A mixture of chloroform, acetone, methanol, water (8:2:2:1) have been shown to be successful mixture for extraction of biological tissues.[118] Extracts are concentrated and resuspended in small volume of appropriate solvent for further identification and quantitation of the test compound. Extracts may require precolumn clean-up prior to analytical identifications. The residue of the skin may be analyzed to determine the bound (unextracted) quantity. Mass balance using the radionuclide is necessary and quantitation is accomplished with the most sensitive analytical technique, either chemical or nuclide.

K. DATA ANALYSIS

Analysis of excretion data for evidence of dermal first-pass metabolism is the most insensitive of the approaches. While the determination of total radioactivity penetrating the skin gives a maximum to the amount of material, parent or product, that is dermally absorbed, it is desirable in some cases to know quantitatively what chemical species are formed in the treated skin area and which species actually penetrates the skin into the systemic circulation. This is a difficult problem presently and all possible techniques that can provide evidence or give even partial information of this topic must be marshalled to solve this problem.

Skin homogenates, explant cultures, and isolated skin flaps may be dosed to determine what products or metabolites might be produced in the *in vivo* situation. Special studies, depending on the chemical characteristics of the compound may need to be considered. These experiments provide evidence suggesting that the same metabolism may occur *in vivo*.

Unfortunately the determination of the occurrence of dermal metabolism *in vivo* is perhaps even more difficult to unequivocally substantiate. Confounding factors such as concurrent metabolism in blood or other tissues of the body and nonlinear kinetics of metabolism and organ distribution may impede interpretation of experimental results between bolus dosing and the somewhat extended dermal penetration. However, various techniques are sometimes applicable in particular situations and for various assumptions regarding the physiological pharmacokinetics of a specific compound.

Comparison of the parent and metabolite(s) or total excretion to either parent or metabolite following intravenous and dermal exposure can sometimes suggest dermal metabolism *in vivo*. Even if this is theoretically possible, the biological, experimental, and analytical variability may preclude identification. It is also necessary to match the dose used for iv and dermal exposure as well as their time course of delivery into the systemic circulation. Nonlinear pharmacokinetics can give false indications of dermal metabolism when only a magnitude difference in the dose is the real cause. Tracer doses are therefore indicated, placing a burden on the analytic sensitivity.

Dermal metabolism may sometimes be inferred from analysis of the blood for parent and metabolites following dermal and intravenous dosing. The advantage of blood analysis rather than excreta is that blood is the proximate tissue for the dermal penetrant. The disadvantage is that blood levels can be very low and decrease rapidly to nondetectable levels.

Use of a pharmacokinetic model may help in answering the question of dermal metabolism. A physiological pharmacokinetic model[120-126] can a priori be developed to describe the kinetics of the parent and metabolites after intravenous administration. The more organs and tissues assayed for the compound and its metabolites, the more unique the model and the more sensitive the analysis will be in detecting first pass metabolism occurring in skin. Use of a tracer dose will permit calculation of the transfer coefficients under linear conditions. At least two higher doses are necessary to determine the enzymatic structured constants. This information from the intravenous experiment fixes the model for the behavior of the chemical after appearance in the blood. Pharmacokinetic analysis using data from the same organs and tissues at various times after dermal application is then limited to those parameters associated with the dermal penetration processes. Physiological pharmacokinetic models have proved to be quite useful in determining the effect(s) of the body on the chemical and for the extrapolation of results from one species to another.

1. Commentary

The identification and quantitation of first-pass metabolism in the skin is a major undertaking that will require information from both *in vivo* and *in vitro* experiments. It is a challenge for the analytical chemist, the biologist, and the kineticist. It is important to recognize that because of the possible nonlinearities in disposition, the parenteral dose and the dermal penetrated dose should be similar. Consideration should be given to the use of infusion studies of parent chemical, major metabolites, and biologically active metabolites as infusion studies can more closely simulate the percutaneous absorption kinetics than can bolus administration. The use of radioactively labeled compounds in metabolism studies facilitates both the quantitative and qualitative aspects of the study (mass balance).

ACKNOWLEDGMENT

The authors acknowledge the learned counsel of their colleague and friend Henry L. Fisher, Ph.D. in the preparation of this chapter.

REFERENCES

1. **Bronaugh, R. L. and Maibach, H. I.,** Eds., *Percutaneous Absorption,* Marcel Dekker, New York, 1985.
2. **Marzulli, F. N. and Maibach, H. I.,** Eds., *Dermatotoxicology,* Hemisphere, New York, 1987.
3. **Barry, B. W.,** *Dermatological Formulations: Percutaneous Absorption,* Marcel Dekker, New York, 1983.
4. **Schaefer, H., Zesch, A., and Stuttgen, G.,** *Skin Permeability,* Springer-Verlag, New York, 1982.
5. **Bartek, M. J. and LaBudde, J. A.,** Percutaneous absorption, in vivo, in *Animal Models in Dermatology,* Maibach, H. I., Ed., Churchill-Livingstone, New York, 1975, 103.
6. **Cooper, E. R. and Berner, B.,** Skin permeability, in *Methods in Skin Research,* ed. by Skerrow, D. and Skerrow, C. J., John Wiley and Sons, 1985, 407.
7. **Idson, B.,** Percutaneous absorption, *J. Pharm. Sci.,* 64(6), 901, 1975.
8. **Marzulli, F. N., Brown, D. W. C., and Maibach, H. I.,** Techniques for studying skin penetration, *Toxicol. Appl. Pharmacol.,* Suppl. No. 3, 76, 1969.
9. **Nugent, F. J. and Wood, J. A.,** Methods for the study of percutaneous absorption, *Can. J. Pharm. Sci.,* 15, 1, 1980.
10. **Stoughton, R. B.,** Some in vivo and in vitro methods for measuring percutaneous absorption, in Prog. Biol. Sci. Relation Dermatol., 2, 263, 1964.
11. **Grasso, P. and Lansdown, A. B. G.,** Methods of measuring, and factors affecting, percutaneous absorption, *J. Soc. Cosmet. Chem.,* 23, 481, 1972.
12. **Wahlberg, J. E.,** Percutaneous absorption, *Curr. Probl. Dermatol.,* 5, 1, 1973.
13. **Baker, H. and Kligman, A. M.,** A simple in vivo method for studying the permeability of the human stratum corneum, *J. Invest. Dermatol.,* 48, 273, 1967.
14. **Petersen, R. V., Kislalioglu, M. S., Liang, W-Q., Fang, S-M., Emam, M., and Dickman, S.,** The athymic nude mouse grafted with human skin as a model for evaluating the safety and effectiveness of radiolabeled cosmetic ingredients, *J. Soc. Cosmet. Chem.,* 37, 249, 1986.
15. **MacKee, G. M., Sulzberger, M. B., Herrmann, F., and Baer, R. L.,** Histologic studies on percutaneous penetration with special reference to the effect of vehicles, *J. Invest. Dermatol.,* 6, 43, 1945.
16. **Rutherford, T. and Black, J. G.,** The use of autoradiography to study the localization of germicides in skin, *Br. J. Dermatol.,* 81, Suppl. 4, 75, 1969.
17. **Malkinson, F. D.,** Studies on the percutaneous absorption of ^{14}C labeled steroids by use of the gas-flow cell, *J. Invest. Dermatol.,* 31, 19, 1958.

18. **Ainsworth, M.,** Methods for measuring percutaneous absorption, *J. Soc. Cosmet. Chem.,* 11, 69, 1960.

19. **Wahlberg, J. E.,** Disappearance measurements, a method for studying percutaneous absorption of isotope-labelled compounds emitting gamma-rays, *Acta Derm Venereol.,* 45, 397, 1965.

20. **Hunziker, N., Feldmann, R. J., and Maibach, H. I.,** Animal models of percutaneous penetration: comparison between mexican hairless dogs and man, *Dermatologica,* 156, 79, 1978.

21. **Anjo, D. M., Feldmann, R. J., and Maibach, H. I.,** Methods for predicting percutaneous penetration in man, in *Percutaneous Absorption of Steroids,* Mauvais-Jarvis, P., Vickers, C. F. H., and Wepierre, J., Eds., Academic Press, New York, 1980, 31.

22. **Ullberg, S. and Appelgren, L-E.,** Experiences in locating drugs at different levels of resolution, in *Autoradiography of Diffusible Substances,* Roth, L. J. and Stumpf, W. E., Eds., Academic Press, New York, 1969, 279.

23. **Ugarte, A. S., Colombetti, L. G., Mewissen, D. J.,** Tissue preparation for autoradiography: the autoradiographic process, in *Principles of Radiopharmacology,* Vol. 1, Colombetti, L. G., Ed., CRC Press, Boca Raton, FL, 1979, 87.

24. **Macht, D. I.,** The absorption of drugs and poisons through the skin and mucous membranes, *JAMA,* 110, 409, 1938.

25. **Brown, V. K.,** *Acute Toxicity in Theory and Practice,* J. Wiley and Sons, New York, 1980, 94.

26. **Gaines, T. B.,** The acute toxicity of pesticides to rats, *Toxicol. Appl. Pharmacol.,* 2, 88, 1960.

27. **Gaines, T. B.,** Acute toxicity of pesticides, *Toxicol. Appl. Pharmacol.,* 14, 515, 1969.

28. **Gaines, T. B. and Linder, R. E.,** Acute toxicity of pesticides in adult and weanling rats, *Fund. Appl. Toxicol.,* 7, 299, 1986.

29. **Noakes, D. N. and Sanderson, D. M.,** A method for determining the dermal toxicity of pesticides, *Br. J. Ind. Med.,* 26, 59, 1969.

30. **Baker, E. L., Jr., Zack, M., Miles, J. M., Alderman, L., Warren, M., Dobbin, R. D., Miller, S., and Teeters, W. R.,** Epidemic malathion poisoning in Pakistan malaria workers, *Lancet,* 1, 31, 1978.

31. **Meier, E. P., Dennis, W. H., Rosencrance, A. B., Randall, W. J., Cooper, W. J., and Warner, M. C.,** Sulfotepp, a toxic impurity in formulations of diazinon, *Bull. Environ. Contam. Toxicol.,* 23, 158, 1979.

32. **Draize, J. H.,** Dermal toxicity, in *Association of Food and Drug Officials of the United States,* Austin, Texas. Appraised of the Safety of Chemicals in Food, Drug and Cosmetics, 1959, 46.

33. **Deichmann, W. B. and LeBlanc, T. J.,** Determination of the approximate lethal dose with about six animals, *J. Ind. Hyg. Toxicol.,* 25, 415, 1943.

34. **Hayes, W. J., Jr.,** *Pesticides Studied in Man,* Williams and Wilkins, Baltimore, 1982, 304.

35. **Knaak, J. B., Yee, K., Ackerman, C. R., Zweig, G., Fry, D. M., and Wilson, B. W.,** Percutaneous absorption and dermal dose-cholinesterase response studies with parathion and carbaryl in the rat, *Toxicol. Appl. Pharmacol.,* 76, 252, 1984.

36. **Shivanandappa, T., Joseph, P., and Krishnakumari, M. K.,** Response of blood and brain cholinesterase to dermal exposure of bromophos in the rat, *Toxicology,* 48, 199, 1988.

37. **Knaak, J. B., Maddy, K. T., Jackson, T., Fredrickson, A. S., Peoples, S. A., and Love, R.,** Cholinesterase activity in blood samples collected from field workers and nonfield workers in California, *Toxicol. Appl. Pharmacol.,* 45, 755, 1978.

38. **Ellman, G. L., Courtney, K. D., Andres, V., Jr., and Featherstone, R. M.,** A new and rapid colorimetric determination of acetylcholinesterase activity, *Biochem. Pharmacol.,* 7, 88, 1961.

39. **U.S. Environmental Protection Agency,** Cholinesterase Activity in Blood, Manual of Analytical Methods for the Analysis of Pesticides in Humans and Environmental Samples, EPA-600/8-80-038, Sec.6,A,(3),(a), 1980, 1,.

40. **Skinner, C. S. and Kilgore, W. W.,** Application of a dermal selfexposure model to worker reentry, *J. Toxicol. Environ. Health,* 9, 461, 1982.

41. **Haleblian, J. K.,** Bioassays used in development of topical dosage forms, *J. Pharm. Sci.,* 65, 1417, 1976.

42. **Shelley, W. B. and Melton, F. M.,** Factors accelerating the penetration of histamine through normal intact human skin, *J. Invest. Dermatol.,* 13, 61, 1949.

43. **Nachman, R. L. and Esterly, N. B.,** Increased skin permeability in preterm infants, *J. Pediatr.,* 79, 628, 1971.

44. **Albery, W. J., Guy, R. H., and Hadgraft, J.,** Percutaneous absorption: transport in the dermis, *Int. J. Pharmaceut.,* 15, 125, 1983.

45. **Nilsson, G. E., Otto, U., and Wahlberg, J. E.,** Assessment of skin irritancy in man by laser doppler flowmetry, *Contact Dermat.,* 8, 401, 1982.

46. **Ryatt, K. S., Stevenson, J. M., Maibach, H. I., and Guy, R. H.,** Pharmacodynamic measurement of percutaneous penetration enhancement in vivo, *J. Pharm. Sci.,* 75, 374, 1986.

47. **Guy, R. H., Tur, E., Bugatto, B., Gaebel, C., Sheiner, L. B., and Maibach, H. I.,** Pharmacodynamic measurements of methyl nicotinate percutaneous absorption, *Pharmaceut. Res.,* 1, 76, 1984.

48. **Drouard, V., Wilson, D. R., Maibach, H. I., and Guy, R. H.,** Quantitative assessment of UV-induced changes in microcirculatory flow by laser doppler velocimetry, *J. Invest. Dermatol.,* 83, 188, 1984.
49. **Amantea, M., Tur, E., Maibach, H. I., and Guy, R. H.,** Preliminary skin blood flow measurements appear unsuccessful for assessing topical corticosteroid effect, *Arch. Dermatol. Res.,* 275, 419, 1983.
50. **McKenzie, A. W. and Stoughton, R. B.,** Method for comparing percutaneous absorption of steroids, *Arch. Dermatol.,* 86, 608, 1962.
51. **Kirsch, J., Gibson, J. R., Darley, C. R., and Burke, C. A.,** A comparison of the potencies of several diluted and undiluted corticosteroid preparations using the vasoconstrictor assay, *Dermatologica,* 167, 138, 1983.
52. **Woodford, R. and Barry, B. W.,** Comparative bioavailability of proprietary hydrophilic topical steroid preparations, *J. Pharm. Pharmacol.,* 25, Suppl., 123, 1973.
53. **Barry, B. W. and Woodford, R.,** Comparative bioavailability and activity of proprietary topical corticosteroid preparations: vasoconstrictor assays on thirty-one ointments, *Br. J. Dermatol.,* 93, 563, 1975.
54. **Litchfield, J. T., Jr. and Wilcoxon, F.,** A simplified method of evaluating dose-effect experiments, *J. Pharmacol. Exp. Ther.,* 96, 99, 1949.
55. **Finney, D. J.,** Probit Analysis, 3rd ed., Cambridge Press, London, 1971.
56. **SAS Institute, Inc.,** SAS User's Guide: Statistics, proc probit, Version 5 Edition, Cary, NC.
57. **Loden, M.,** Methods of detection, in *Methods for Skin Absorption,* Kemppainen, B. W. and Reifenrath, W. G., Eds., CRC Press, Boca Raton, FL, 1989, in press.
58. **Hodge, H. C. and Sterner, J. H.,** The skin absorption of triorthocresyl phosphate as shown by radioactive phosphorus, *J. Pharm. Exp. Ther.,* 79, 225, 1943.
59. **Cocchetto, D. M. and Bjornsson, T. D.,** Methods for vascular access and collection of body fluids from the laboratory rat, *J. Pharm. Sci.,* 72, 465, 1983.
60. **Flynn, L. A. and Guilloud, R. B.,** Vascular catheterization: advantages over venipuncture for multiple blood collection, *Lab. Anim.,* 29, 1988.
61. **Hall, L. L., DeLopez, O. H., Roberts, A., and Smith, F. A.,** A procedure for chronic intravenous catheterization in the rabbit, *Lab. Anim. Sci.,* 24, 79, 1974.
62. **Shah, P. V., Fisher, H. L., Month, N. J., Sumler, M. R., and Hall, L. L.,** Dermal penetration of carbofuran in young and adult fischer 344 rats, *J. Toxicol. Environ. Health,* 22, 207, 1987.
63. **Wester, R. C., Noonan, P. K., Smeach, S., and Kosobud, L.,** Pharmacokinetics and bioavailability of intravenous and topical nitroglycerin in the rhesus monkey: estimate of percutaneous first-pass metabolism, *J. Pharm. Sci.,* 72, 745, 1983.
64. **McDougal, J. N., Jepson, G. W., Clewell, H. J., III, and Andersen, M. E.,** Dermal absorption of dihalomethane vapors, *Toxicol. Appl. Pharmacol.,* 79, 150, 1985.
65. **Ogiso, T., Ito, Y., Iwaki, M., and Atago, H.,** Prediction of plasma concentration profile during single and repeated skin applications of indomethacin and its calcium salt ointments, *J. Pharmacobio Dyn.,* 10, 384, 1987.
66. **Ogiso, T., Ito, Y., Iwaki, M., Yamamoto, Y., and Yamahata, T.,** Percutaneous absorption of valproic acid and its plasma concentration after application of ointment, *J. Pharmacobio Dyn.,* 10, 537, 1987.
67. **Ogiso, T., Ito, Y., Iwaki, M., Atago, H., Tanaka, C., Maniwa, N., and Ishida, S.,** Percutaneous absorption of dexamethasone acetate and palmitate, and the plasma concentration, *Chem. Pharm. Bull.(Tokyo),* 35, 4263, 1987.
68. **Ogiso, T., Ito, Y., Iwaki, M., and Shintani, A.,** The percutaneous absorption of propranolol and prediction of the plasma concentration, *J. Pharmacobio Dyn.,* 11, 349, 1988.
69. **Ogiso, T., Ito, Y., Iwaki, M., Atago, H., and Yamamoto, Y.,** A pharmacokinetic model for percutaneous absorption of valproic acid and prediction of drug disposition, *J. Pharmacobio Dyn.,* 11, 444, 1988.
70. **Kondo, S., Mizuno, T., and Sugimoto, I.,** Effects of penetration enhancers on percutaneous absorption of nifedipine. Comparison between deet and azone, *J. Pharmacobio Dyn.,* 11, 88, 1988.
71. **Hwang, C-C. and Danti, A. G.,** Percutaneous absorption of flufenamic acid in rabbits: effect of dimethyl sulfoxide and various nonionic surface-active agents, *J. Pharm. Sci.,* 72, 857, 1983.
72. **Poiger, H. and Schlatter, Ch.,** Influence of solvents and adsorbents on dermal and intestinal absorption of TCDD, *Food Cosmet. Toxicol.,* 18, 477, 1980.
73. **Laug, E. P. and Kunze, F. M.,** The penetration of lead through the skin, *J. Ind. Hyg. Toxicol.,* 30, 256, 1948.
74. **White, W. J. and Field, K. J.,** Anesthesia and surgery of laboratory animals, *Vet. Clin. North Am. (Small Anim. Pract.),* 17, 989, 1987.
75. **Fisher, H. L., Most, B., and Hall, L. L.,** Dermal absorption of pesticides calculated by deconvolution, *J. Appl. Toxicol.,* 5, 163, 1985.
76. **Sato, K., Oda, T., Sugibayashi, K., and Morimoto, Y.,** Estimation of blood concentration of drugs after topical application from in vitro skin permeation data. I. Prediction by convolution and confirmation by deconvolution, *Chem. Pharm. Bull. (Tokyo),* 36, 2232, 1988.

77. **Smolen, V. F.,** A generalized numerical deconvolution procedure for computing absolute bioavailability — time profiles, in *Kinetic Data Analysis: Design and Analysis of Enzyme and Pharmacokinetic Experiments,* Endrenyi, L., Ed., 1981, 375.

78. **Guy, R. H. and Hadgraft, J.,** Pharmacokinetic interpretation of the plasma levels of clonidine following transdermal delivery, *J. Pharm. Sci.,* 74, 1016, 1985.

79. **Berman, M. and Weiss, M. F.,** SAAM Manual, DHEW Publication No,(NIH) 78, 1978.

80. **Feldmann, R. J. and Maibach, H. I.,** Percutaneous penetration of steroids in man, *J. Invest. Dermatol.,* 52, 89, 1969.

81. **Feldmann, R. J. and Maibach, H. I.,** Absorption of some organic compounds through the skin in man, *J. Invest. Dermatol.,* 54, 399, 1970.

82. **Bucks, D. A. W., Maibach, H. I., and Guy, R. H.,** Percutaneous absorption of steroids: effect of repeated application, *J. Pharm. Sci.,* 74, 1337, 1985.

83. **Bartek, M. J., LaBudde, J. A., and Maibach, H. I.,** Skin permeability in vivo: comparison in rat, rabbit, pig, and man, *J. Invest. Dermatol.,* 58, 114, 1972.

84. **Bucks, D. A. W., Marty, J. P. L., and Maibach, H. I.,** Percutaneous absorption of malathion in the guinea pig: effect of repeated topical application, *Food Chem. Toxicol.,* 23, 919, 1985.

85. **Wester, R. C., Bucks, D. A. W., Maibach, H. I., and Anderson, J.,** Polychlorinated biphenyls(PCBs): dermal absorption, systemic elimination, and dermal wash efficiency, *J. Toxicol. Environ. Health,* 12, 511, 1983.

86. **Hunziker, N., Feldmann, R. J., and Maibach, H. I.,** Animal models of percutaneous penetration: comparison between mexican hairless dogs and man, *Dermatologica,* 156, 79, 1978.

87. **Andersen, K. E., Maibach, H. I., and Anjo, M. D.,** The guinea-pig: an animal model for human skin absorption of hydrocortisone, testosterone and benzoic acid, *Br. J. Dermatol.,* 102, 447, 1980.

88. **Ando, H. Y., Sugita, E. T., Schnaare, R. L., and Bogdanowich, L.,** Guinea pig ear as a new model for in vivo percutaneous absorption, *J. Pharm. Sci.,* 71, 1157, 1982.

89. **Bucks, D. A. W., McMaster, J. R., Maibach, H. I., and Guy, R. H.,** Bioavailability of topically administered steroids: a "mass balance" technique, *J. Invest. Dermatol.,* 90, 29, 1988.

90. **Shah, P. V., Fisher, H. L., Sumler, M. R., Monroe, R. J., Chernoff, N., and Hall, L. L.,** Comparison of the penetration of 14 pesticides through the skin of young and adult rats, *J. Toxicol. Environ. Health,* 21, 353, 1987.

91. **Bronaugh, R. L., Stewart, R. F., Congdon, E. R., and Giles, A. L.,** Methods for in vitro percutaneous absorption studies. I. Comparison with in vivo results, *Toxicol. Appl. Pharmacol.,* 62, 474, 1982.

92. **Guy, R. H., Hadgraft, J., and Maibach, H. I.,** Percutaneous absorption in man: a kinetic approach, *Toxicol. Appl. Pharmacol.,* 78, 123, 1985.

93. **Marco, G. J., Simoneaux, B. J., Williams, S. C., Cassidy, J. E., Bissig, R., and Muecke, W.,** Radiotracer approaches to rodent dermal studies, in *Dermal Exposure Related to Pesticide Use,* Honeycutt, R. C., Zweig, D., and Ragsdale, N., Eds., ACS Symposium Series, No. 273, 1985, 44.

94. **Shah, P. V. and Guthrie, F. E.,** Dermal absorption, distribution, and the fate of six pesticides in the rabbit, in *Pesticide Managment and Insecticide Resistance,* Watson, D. L. and Brown, A. W. A., Eds., Academic Press, New York, 1977, 547.

95. **Shah, P. V. and Guthrie, F. E.,** Percutaneous penetration of three insecticides in rats: a comparison of two methods for in vivo determination, *J. Invest. Dermatol.,* 80, 291, 1983.

96. **Reifenrath, W. G., Hill, J. A., Robinson, P. B., McVey, D. L., Akers, W. A., Anjo, D. M., and Maibach,, H. I.,** Percutaneous absorption of carbon 14 labeled insect repellents in hairless dogs, *J. Environ. Pathol. Toxicol.,* 4, 249, 1980.

97. **Reifenrath, W. G., Chellquist, E. M., Shipwash, E. A., Jederberg, W. W., and Krueger, G. G.,** Percutaneous penetration in the hairless dog, weanling pig, and grafted athymic nude mouse: evaluation of models for predicting skin penetration in man, *Br. J. Dermatol.,* 111, Suppl. 27, 123, 1984.

98. **Reifenrath, W. G., Chellquist, E. M., Shipwash, E. A., and Jederberg, W. W.,** Evaluation of animal models for predicting skin penetration in man, *Fund. Appl. Toxicol.,* 4, S224-S230, 1984.

99. **Black, K. E. and Jederberg, W. W.,** Athymic nude mice and human skin grafting, in *Models in Dermatology,* Vol. 1, Maibach, H. I. and Lowe, Eds., Karger, Basel, 1985, 228.

100. **Chow, C., Chow, A. Y. K., Downie, R. H., and Buttar, H. S.,** Percutaneous absorption of hexachlorophene in rats, guinea pigs, and pigs, *Toxciology,* 9, 147, 1978.

101. **Bronaugh, R. L., Stewart, R. F., and Congdon, E. R.,** Differences in permeability of rat skin related to sex and body site, *J. Soc. Cosmet. Chem.,* 34, 127, 1983.

102. **Shah, P. V., Monroe, R. J., and Guthrie, F. E.,** Comparative penetration of insecticides in target and non-target species, *Drug Chem. Toxicol.,* 6, 155, 1983.

103. **Olson, W. P. and O'Brien, R. D.,** The relation between physical properties and the penetration of solutes into the cockroach cuticle, *J. Insect Physiol.,* 9, 777, 1963.

104. **Buerger, A. A. and O'Brien, R. D.,** Penetration of nonelectrolytes through animal integuments, *J. Cell. Comp. Physiol.,* 66, 227, 1965.

105. **Abou-Donia, M. B.**, Neurotoxicity produced by long-term low-level percutaneous administration of leptophos, in *Proceedings of 1st International Congress on Toxicology,* Tollefson Lithographing, Oakville, Ontario, Canada, 1977.

106. **Glees, P.**, A morphological and neurological analysis of neurotoxicity illustrated by tricresyl phosphate intoxication in the chick, in *Neurotoxicity of Drugs, Proceedings of the European Society for the Study of Drug Toxicity,* Davey, G., Ed., Vol. 8, 1967, 136.

107. **Krishnamurti, A., Kanagasuntheram, R., and Vij, S.**, Effect of TOCP poisoning on the pacinian corpuscles of slow loris, *Acta Neuropathol.,* 22, 345, 1972.

108. **Abou-Donia, M. B.**, Pharmacokinetics and metabolism of a topically applied dose of O-4-bromo-2,5-dichlorophenyl O-methyl phenylphosphonothioate in hens, *Toxicol. Appl. Pharmacol.,* 51, 311, 1979.

109. **Noonan, P. K. and Wester, R. C.**, Cutaneous metabolism of xenobiotics, in *Percutaneous Absorption: Mechanisms, Methodology, and Drug Delivery,* Bronaugh, R. L. and Maibach, H. I., Eds., 1985, 65.

110. **Noonan, P. K. and Wester, R. C.**, Cutaneous biotransformation: some pharmacological and toxicological implications, in *Dermatotoxicology,* 3rd ed., Marzulli, F. N. and Maibach, H. I., Eds., Hemisphere, New York, 1987, chap. 2.

111. **Weston, A., Grover, P. L., and Sims, P.**, Metabolism and activation of benzo[a]pyrene by mouse and rat skin in short-term organ culture and in vivo, *Chem. Biol. Interact.,* 42, 233, 1982.

112. **Coomes, M. W., Norling, A. H., Pohl, J., Muller, D., and Fouts, J. R.**, Foreign compound metabolism by isolated skin cells from the hairless mouse, *J. Pharmacol. Exp. Ther.,* 225, 770, 1983.

113. **Kao, J., Patterson, F. K., and Hall, J.**, Skin penetration and metabolism of topically applied chemicals in six mammalian species, including man: an in vitro study with benzo[a]pyrene and testosterone, *Toxicol. Appl. Pharmacol.,* 81, 502, 1985.

114. **Segal, A., Van Durren, B., and Mate, U.**, The identification of phorbol myristate acetate as a new metabolite of phorbol myristate acetate in mouse skin, *Cancer Res.,* 35, 2154, 1975.

115. **Lukas, G., Brindle, S. D., and Greengard, P.**, The route of absorption of intraperitoneally administered compounds, *J. Pharmacol. Exp. Ther.,* 178, 562, 1971.

116. **Renwick, A. G., Pettet, J. L., Grouchy, B., and Corina, D. L.**, The fate of 14-C-trithiozine in man, *Xenobiotica,* 12, 329, 1982.

117. **Hoffman, K-J., Arfwidsson, A., and Borg, K. O.**, The metabolic disposition of the selective beta(1)-adrenoceptor agonist prenalterol in mice, rats, dogs, and humans, *Drug. Metab. Disp.,* 10, 173, 1982.

118. **Ioannou, Y. M. and Matthews, H. B.**, Absorption, distribution, metabolism, and excretion of 1,3-diphenylguanidine in the male Fischer 344 rat, *Fund. Appl. Toxicol.,* 4, 22, 1984.

119. **Shah, P. V., Sumler, M. R., Ioannou, Y. M., Fisher, H. L., and Hall, L. L.**, Dermal absorption and disposition of 1,3-diphenylguanidine in rats, *J. Toxicol. Environ. Health,* 15, 623, 1985.

120. *Pharmacokinetics in Risk Assessment: Drinking Water and Health,* Natl. Acad. Press, Washington, D.C., 1987, 8.

121. **Pecile, A. and Rescigno, A., Eds.,** *Pharmacokinetics: Mathematical and Statistical Approaches to Metabolism and Distribution of Chemicals and Drugs,* Plenum Press, New York, 1988.

122. **D'Souza, R. W. and Andersen, M. E.**, Physiologically based pharmacokinetic model for vinylidene chloride, *Toxicol. Appl. Pharmacol.,* 95, 230, 1988.

123. **McDougal, J. N., Jepson, G. W., Clewell, H. J., III, MacNaughton, M. G., and Andersen, M. E.**, A physiological pharmacokinetic model for dermal absorption of vapors in the rat, *Toxicol. Appl. Pharmacol.,* 85, 286, 1986.

124. **Gerlowski, L. E. and Jain, R. K.**, Physiologically based pharmacokinetic modeling: principles and applications, *J. Pharm. Sci.,* 72, 1103, 1983.

125. **Himmelstein, K. J. and Lutz, R. J.**, A review of the applications of physiologically based pharmacokinetic modeling, *J. Pharmacokin. Biopharm.,* 7, 127, 1979.

126. **Bischoff, K. B., Dedrick, R. L., and Zaharko, D. S.**, Preliminary model for methotrexate pharmacokinetics, *J. Pharm. Sci.,* 59, 149, 1970.

Chapter 7

PHYSICAL RESOURCES NEEDED FOR *IN VIVO* ANIMAL STUDIES*

Hubert L. Snodgrass

TABLE OF CONTENTS

---

* Use of trademarked names does not imply endorsement by the U.S. Army, but is intended only to assist in the identification of a specific product. The opinions and assertions contained herein are the private views of the author and are not to be construed as official or as reflecting views of the Department of the Army or the Department of Defense (AR 360-5).

I. INTRODUCTION

The goal of the *in vivo* percutaneous penetration study is not to determine how much of a substance goes through the skin of an animal, but rather to provide a basis for predicting how much will be absorbed by man. This holds true for substances intended for human contact, such as cosmetics and drugs, as well as those hazardous materials posing a health risk through accidental exposure. Clearly, the basis for these predictions must rest upon controlled laboratory studies and sound judgement. The intent of this commentary is to keep the investigator in control; to provide rudimentary information on the operation of animal tests which measure dermal absorption and to describe devices and/or equipment needed to that end. Independent of the questions to be answered, the simplicity or complexity of test design, or the political mood surrounding the test substance, the operational side of the animal test remains basically the same. This includes the dermal application of the test substance, protection of the dermally dosed site following treatment, and collection of biological specimens for measurement of absorbed chemical. The experienced animal investigator may remain wary of this simplistic approach. As well he should, for the operation of an *in vivo* skin penetration study in animals makes two protracted assumptions. First, that an animal will in effect wear an extraneous appliance for a period of one day to one month without disturbing it. Second, that the same animal will produce and deposit its excreta in designated areas and not mix it with spilled food or drinking water. Like every football play that was designed to score a touchdown, the odds of animal compliance are about the same. To reduce those odds we have noted some of the avoidable problems; some which have to be lived with.

The remainder of this paper deals mostly with hardware, or what one needs to accomplish a dermal absorption study in animals. The next section applies to those investigators experienced in animal testing but who have not performed *in vivo* percutaneous penetration studies. In addition to study goals, it briefly discusses the use of radioisotopes which are commonly used for the quantitation of skin absorption. These materials are a valuable investigative tool but they also have their own rules. Finally, a manufacturer's index appears at the end of the chapter which, when possible, provides the source of much of the equipment cited.

II. GENERAL CONSIDERATIONS

A. INFORMATION REQUIRED

In the planning stages of the investigation, it must first be determined what information is to be gained. Will the test results be submitted for regulatory review or will they be used only to screen out potentially dangerous substances during the early stages of product development? Also, what resources and technical assets are available? For example, a laboratory of limited size and expertise may only be able to perform a simple quantitation of absorption, perhaps a single application in one animal species. Another organization, having a strong compliment of professional staff may wish to also include compartment kinetics, half-life measurements, and characterization of metabolites following multiple applications to two or more species. It has been our experience that the smaller, well planned and executed project has had far greater impact on health effects projections than ones which were overly ambitious in concept. Start with answering the basic or required questions and build from there.

B. RADIOCHEMICALS

The low energy beta emitters, particularly carbon 14 and tritium, are routinely used as tracers in skin penetration studies. They can be introduced into most organic test substances with relative ease, are readily measurable and present minimal hazard to the investigator. Nonlabeled chemicals as well as stable isotopes are also frequently used but these require sensitive, compound-specific methods of analyses. Many of the more common radiochemicals and drugs are stocked by commercial suppliers. Their limited cost makes them desirable for comparative testing when a particular test system is being developed. Experimental substances, however, must usually be custom synthesized, either within the organization or by a commercial supplier. If the material is to be custom synthesized, the client should be prepared to provide a small quantity of the pure material as a standard, and possibly, the chemical precursors necessary to build the substance. A quotation for synthesis is easily obtained by a phone call to one of the suppliers. The desired site of molecular labeling should be specified as well as the required radiochemical purity. Keep in mind that skin penetration will be quantitated based upon recovery of the radiolabel, usually appearing in the excreta. Therefore, the label (e.g., ^{14}C) should be positioned in the "core" of the compound and follow the major metabolites until excreted. The ^{14}C label should not be metabolically lost as CO_2 or become part of the one-carbon pool of the body.

C. HEALTH PHYSICS

The use of radioisotopes places certain responsbilities upon the new investigator who is often unfamiliar with the area of health physics or the safe handling of radioactive sources. All research organizations licensed by the Nuclear Regulatory Commission (NRC) to possess radioactive sources are required to maintain an internal health physics program. Therefore, investigators contemplating the use of these materials should seek guidance there. Of importance is the identity of isotope, quantity, and proposed use. A responsible investigator will need to be identified which will oversee the use of the controlled materials and assure their safe use. The regulatory requirements may at first appear overwhelming, but like accounting, the initial setup of procedures is the hardest. Once operational, however, an attitude of attentiveness and compliance must be maintained. If the organization is not currently authorized to purchase or possess radioisotopes, the regional offices of the NRC may be contacted for additional guidance. Title 10, Code of Federal Regulations, Chapter I, Nuclear Regulatory Commission (Parts 1-170), is the controlling regulatory document.

D. ANIMAL BEHAVIOR

Anyone who has owned a dog knows that the adolescent is prone to "puddling" when it becomes excited or stressed. To the contrary, laboratory animals usually exhibit the opposite response. We have noted animals that have withheld urine for 3 to 4 d following treatment although they continue to drink. Feces production in rabbits has been delayed for as long as 6 d without known provocation. Since most dermal penetration studies depend entirely upon daily excreta production for the measurement of absorbed chemical, a change in normal output can, and usually does, create havoc when data is collated and averaged. This can be combated in one of two ways, either by increasing the number of animals used (and the number of specimens) to compensate for the outliers or by taking the time to familiarize the selected species to the test routine beforehand. The latter program, if for no other than fiscal reasons, is recommended. We generally restrict access to the animal treatment room to two or three people who are actually going to perform the test operations. A daily routine is established, beginning about 10 days before the test start. During this pretreatment phase, any "problem" animals can be replaced. The intent is to maintain an absolute routine which produces consistent animal behavior and physiological responses.

III. ANIMAL TREATMENT

The topical application of the test substance is usually applied to the mid-lumbar section of the animal's back, the area having been clipped free of hair the day before. In primates, the abdomen or the dorsal surface of the forearm is used. The area to be treated may be demarcated to contain the test substance within a premeasured area, thus assuring a constant exposure rate. We use a section of soldering wire, formed to the desired size. This is dipped into warmed petrolatum and immediate placed on the skin surface. The resulting imprint provides a barrier for containing most liquid substances. The area may also be marked with a felt-tipped pen if the substance can be applied, and stays, within the desired area. The application of the liquid test chemical is performed using a calibrated autopipette or microliter syringe with blunted needle. Dry materials are generally formed into a slurry with water or saline and applied with a spatula. The treatment area for solid materials may be demarcated with a template of adhesive tape or latex dental dam (The Hygenic Corporation, Akron, OH). After the solvent has evaporated (from liquids) the site is covered with a protective device to prevent surface loss of the test material. For rats and mice, light anesthesia is used to facilitate application of the test substance and the protective device. Inhaled anesthetics, such as ether or a 50% CO_2 atmosphere, work well. For the larger species, only minimal restraint is needed. Treatment with additional drugs (e.g., analgesics, diruetics) is not recommended.

IV. PROTECTIVE COVERINGS

The goal of the protective covering which encloses the site of topical application is to prevent the applied substance from reaching the excreta by any route other than by percutaneous absorption. This can occur from the animal probing and ingesting some of the test material, rubbing the skin site against the cage wall, or from the normal exfoliation of skin. It has been demonstrated that within 48 h of topical application to guinea pigs, 53% of an unprotected (uncovered) dose of ^{3}H-inulin was recovered.[1] Of importance was that while 23% of the dose appeared in the urine, 30% was recovered from the cage floor. Conversely, when the treated area was covered (occluded), only 0.2% of the applied dose appeared in the urine and no radioactivity was detected on the cage floor.

The coverings used in most animal studies are of two basic designs, either a flexible type resembling a sponger rubber doughnut or of rigid construction, i.e., a ring or solid cap. If the site is to be occluded (near-airtight), the rigid device is preferred because of its impervious walls. It may then be covered with nonporous materials such as tape, Saran™ wrap (Dow Chemical Co., Midland, MI), or dental dam. The flexible devices are used in non-occlusive applications where air access to the skin surface is to remain unrestricted. In this case, the foam ring is covered with porous materials such as gauze, thin foam sheeting, fine mesh hardware cloth, nylon screen, or some combination thereof.

A. FLEXIBLE

Bartek et al.,[2] in his article comparing absorption kinetics with various animal species, described a nonocclusive foam pad device which has become a benchmark in its simplicity (Figure 1). The device consists of two rectangular pieces of self-adhering foam padding (Reston™, 3M Company, St. Paul, MN). Each has a smaller area from the middle removed, the cutout being larger than the intended area of application. One pad is affixed to the animal's back, then the test material applied. An appropriate amount of time is allowed for evaporation of the solvent. The second pad, with a layer of hardware screen and gauze pad taped to the top, is then placed on top of the first. The device is secured to the animal's back with adhesive tape, wrapped around the front and back of the pad and the trunk of the animal (Figure 2).

Several minor variations to the Bartek design are reported. One modification sandwiches the screen and gauze layers between the two pads;[3] and another method adds a second layer of screen on the outer surface to discourage animals from removing the exposed gauze pad.[4]

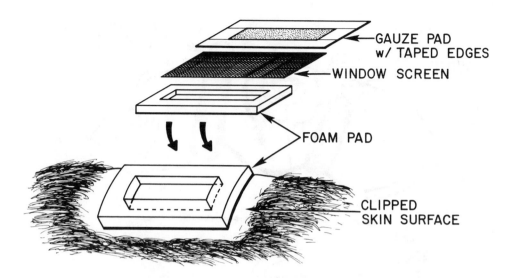

FIGURE 1.   Nonocclusive protective patch described by Bartek et al.

FIGURE 2.   Nonocclusive protective patch following attachment to a mini pig.

The success of the Bartek device results from its flexibility, thus allowing unrestricted mobility of the animal. It works best in the larger, nonrodent animal species, but we have successfully used the device in rats for up to 7 d of continuous wear without major difficulties. In the rat, only a single 0.4 cm thick layer of foam padding is used. A similar design for use in nude mice is constructed of Reston foam padding measuring 2.0 cm × 2.0 cm × 0.4 cm thick.[5] A hole is punched in the middle using a 1.6 cm diameter cork bore. Nylon screen is glued over the hole and a thin layer of surgical gauze inserted into the hole from the inside. The appliance is then secured to the mouse with surgical tape with the cutout centered over the application area.

Another nonocclusive protective covering described for rats and rabbits uses a strip of elastic adhesive bandage (Elastoplast™, Beiersdorf, Inc., Norwalk, Connecticut), 3 in. wide, taped around the animal's trunk.[6] A precut whole in the tape, over the clipped back, determines the exposure area. After application, a section of 260 mesh nylon (Nitex™, Lambert Co., Boston, MA) is placed over the area and a supporting perimeter of foam strips (Molefoam™, Scholl, Inc., Chicago, IL) is affixed. A section of 40 × 50 mesh copper gauze (Jellif Corp., Southport, CT) is then placed on top and the entire appliance taped to the animal using Elastoplast.

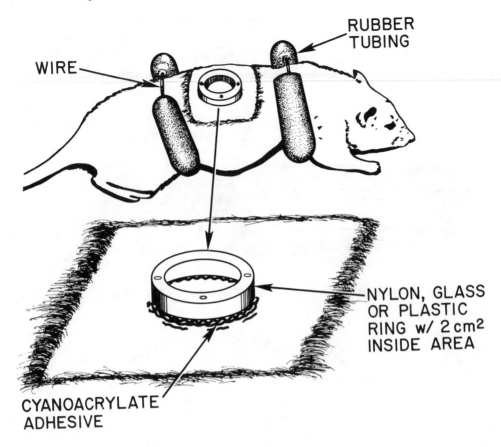

FIGURE 3. Protective device attached to the skin of a rat. Note rubber tubing which prevents interference by the animal.

Flexible coverings which use foam padding or similar materials are generally well tolerated by the animal. We have used the Bartek device in rabbits for exposure periods up to 30 d.[7] That is not to say that some species, particularly the rabbit and the rat, will not probe or attempt to remove them. If rabbits are expected to wear the devices for more than 3 d, an Elizabethan collar (Safe-T-Shield, Ejay International, Glendale, CA) should be considered. A simple collar for rabbits may also be fashioned from a plastic coffee can lid (3 lb size). A whole is cut from the middle approximating the neck size of the rabbit. Another cut is made from the inner circle to the outside to ease fitting of the lid over the rabbit's head. Once in place, the cut is closed with tape. Split latex tubing is glued around the inside cutout to prevent chafing. Elizabethan collars are not recommended for use in rats unless the device will be worn for 24 h or less. The collars inhibit the normal food acquisition instincts in this species and may also prevent drinking. An alternative is to use a section of rubber tubing, about 1 cm in diameter, which encircles the trunk just behind the front legs. This is reported to prevent scratching by the rat and removal of the protective device.[8,9] Bronaugh et al.[10] used two sections of rubber tubing to protect a nylon device being worn by rats (Figure 3). One encircled the body behind the front legs and the other just in front of the rear legs. A short length of wooden dowel, slightly larger than the tubing bore, may be used to join the two ends. Another novel technique for limiting the rat's interference with the protective device is to "shackle" the rear legs of the animal.[11] A short length of jewelry chain terminates at each end with a copper wire loop. The loops encircle each rear leg like handcuffs. The patch is also moved forward to the shoulder area of the back which further restricts accessibility by the rear legs.

B. NONFLEXIBLE

The skin bonding properties of the cyanoacrylate "super glues", while sometimes an unexpected surprise to the general use consumer, may be exploited for the attachment of protective devices to the skin of animals. The greatest advantage is that the appliance may be made of glass or hard plastic which prevents chewing (and ingestion) by the animal. No appreciable skin irritation results from contact with the acrylate glues but removal from the skin should be done with care. It is recommended that the hardened glue be peeled from the skin, not forcibly pulled off.[12] Warm soapy water or baby oil may be used to weaken the skin bond for ease of removal. Silicon rubber cements have also been successfully used to attach plastic or glass devices to the skin surface. Although they must be held in place while the cement dries, usually 2 to 4 min, the bond remains more flexible than that created by the cyanoacrylates. Uncured rubber cements are mildly irritating to the skin. Removal is the same as for the acrylate glues.

The covering device is typically made of a short piece of glass or plastic tubing, roughly 1 to 2 cm long, attached to the skin with glue or cement (Figure 3). The bore size may exceed the desired exposure area or be selected according to a predetermined area. This has significant advantages if same-site, repeated exposures are to be performed. After skin treatment, the open end of the tubing may then be covered with a porous material (non-occlusive) or an impenetrable material (occlusive). A variety of other semi-rigid devices are reported and appear to be limited only by the creativity of the investigator. These include disposable plastic weigh boats (Fisher Scientific, Pittsburgh, PA);[13] a 25 mm Hill Top Chamber (Hill Top Research, Inc., Cincinnati, OH);[13] a plastic blister from a Cathavex single-use filter (Millipore Corp., Bedford, MA) with holes punched in the top with a 20 ga needle;[14] and a 5 ml disposable plastic beaker.[14]

Following the defined exposure period, albeit 8 h or 30 d, the protective covering (either type) is removed and extracted in an appropriate solvent to recover unabsorbed test substance. The unabsorbed material may originate from the evaporated chemical trapped by the appliance or from exfoliated skin at the application site. Foam coverings should be cut into small pieces before extraction. Other materials such as gauze, screen, or tape which could theoretically come in contact with the test chemical are also analyzed. The skin site of application is washed with water or solvent upon removal of the protective covering to recover surface chemical. Care must be taken not to abrade the area or damage the skin with solvents such that the dermal barrier would be compromised.

V. METABOLISM CAGES

Metabolism cages are designed for the purpose of separating animal excreta (urine and feces) such that metabolically eliminated substances can be measured. More sophisticated (closed) systems also provide for measuring expired respiratory gases, usually CO_2. Recent design modifications by caging manufacturers, particularly to the rodent size cage, have eliminated most of the problems inherent to the earlier models. These were the separation of excreta such that urine did not wash over feces, and the prevention of food particles from reaching excreta collection vessels. Complete separation of urine and feces is important for the quantitative identification of bioelimination pathways, and for the isolation of excreted metabolites. The effects of cross contamination have less impact on mass balance studies wherein accountability of the test compound (from any source) is the greater issue. Similarly, food debris mixed with excreta often compromises sensitive chemical tests and also encourages bacterial growth. Excess drinking water spillage can markedly dilute urine specimens.

A. SIZE

The first consideration of metabolism cage design is the overall internal size. This will

obviously vary depending on the species to be housed but must meet the minimum requirements established in DHHS Publication No. (NIH) 85-23. Caging manufacturers, as a rule,
comply with these standards. However, if custom designs are to be used the correct dimensions must be specified by the client. Earlier metabolism cages often do not meet the
current standards and should be replaced or modified accordingly.

B. EXCRETA SEPARATION

Separation of urine and feces, in its simplest form, is accomplished by placing a screen
beneath the cage floor to trap fecal pellets. A tapered, or funnel shaped pan under the screen
then collects the urine into a central collection vessel. This arrangement is still used in most
large animal metabolism cages, e.g., rabbit, dog, pig and monkey, as well as the older
rodent cages. The major fault of this design is that urine usually washes over the feces and
that the taper of the urine collection pan is too shallow for complete runoff. Accordingly,
two modifications are suggested. The screen beneath the cage floor should be tapered
upwards, forming a pyramid of about 6 in. at the peak (for a large animal cage). A lip
around the periphery of the base, extending upwards, should also be included. This design
has the effect of diverting fecal pellets to the sides of the cage screen thus reducing the
chances of urine contamination. The screen should be rigid (not window screen) and have
about a $1/_4$ in. mesh. The urine collection pan needs to have a positive drop of about 4 in.
to the center collection hole, or greater if the collection hole is at one end. Rabbit urine,
being more viscous than from other animals, tends to pool in the collection pan unless an
extreme downward taper is maintained. A Teflon® (E. I. DuPont de Nemours & Co.,
Wilmington, DE) or silicon coating of the pan is suggested. Residual urine should be removed
from the pan at collection milestones with a small amount of distilled water or solvent to
assure complete accountability of the test substance.

The smaller rodent metabolism cages have attained a degree of sophistication not shared
with their larger counterparts. Separation of urine and feces is nearly assured in the new
designs and food and water containment is greatly improved. A wash down with water or
solvent is, however, still suggested at collection intervals to remove any residue. These next-
generation cages (e.g., Nalge) are available in varying sizes for mice to rabbits.

Primates are restrained in a metabolism chair through the first few hours after dermal
exposure. This is the only practical means of preventing hand (or foot) contact with the test
site. Application is made to the clipped abdominal area or the midventral forearm and covered
with either a nonocclusive or impervious device. After the exposure period, primates are
quartered in metabolism cages for collection of excreta for the remainder of the study period.

In all cases, excreta collection must be performed at least once per day and the screen,
trays and collection vessels washed thoroughly. This discourages bacterial growth which
can metabolically degrade the substances. A weak acid solution may also be added to each
fresh urine container to reduce bacterial growth and to stabilize excreted chemicals. We
have also used toluene as a "floating cap" in urine collection vessels to control evaporation.
In this case, as the collection vessel fills, the layer of toluene (0.25 cm) will always partition
above the urine surface. The floating cap is recommended for small containers with limited
surface areas but may also be used in closed systems where the direction of airflow moves
away from the animal chamber. Cooling of the collection vessel by one of any number of
methods is advisable when quantitative assessments of metabolites are required. This limits
evaporation of volatile moieties and tends to stabilize reactive substances.

When rabbits are used, hair should be removed from the screen before feces collection.
Either brush it away or remove it with a hand-held vaccum cleaner. When feces are homogenized in a blender for later analysis, the shed hair invariably clogs the blades. Maintaining the rabbit holding room at 68°F greatly reduces the volume of shed hair.

C. FOOD/WATER ISOLATION

The isolation of drinking water and food from the excreta collection area remains a problem, even among the best cage designs. If a short term test (6 h or less) is to be performed, food may be withheld. Longer periods, however, require some accommodation. The current rodent cage designs (e.g., Nalge and Harvard) isolate the food and water reservoirs from the main living area. The containers are placed at the end of a feeding chamber or short tunnel. Rodents, however, routinely collect food blocks from the open containers and accumulate them in the living area; the worst possible location. This can be minimized by providing ground food although some food dust will always find its way to the living area, clinging to rat's hair and paws. Another approach is to mix a raw egg with the pulverized feed, shaping it to the desired size, then baking it. This has shown some promise in eliminating the dust problem and in forming an immovable food source. Gelled and/or liquid diets are available for rodents and rabbits and may be the answer to preventing food contamination of urine and feces. The gelled diets have been successfully used in rodent shipping containers. Intuitively, animals receiving a liquid diet would need to be conditioned to this change from solid food well in advance of a planned study. The potential for metabolic changes resulting from a liquid diet should also be questioned.

Rabbits, dogs and pigs are less possessive about their food supply but much more wasteful. Accordingly, design improvements focus on deepening the food reservoir such that spillage is minimized. It is also preferrable to keep the food container outside the cage. Dogs in particular tend to "paddle" their front feet in door mounted food and water pans when they become excited, scattering both throughout the cage. A cutout in the cage wall through which the animal extends its head to eat is effective in keeping unwanted food or water from the excreta collection vessels. Dogs and pigs should be fed at specific intervals to control excessive food spills. Feedings at the end of the day or when room operations have been completed appears to minimize distractions.

As a last resort, most food residue can be filtered from urine before it reaches the collection vessel by placing a wad of glass wool at the base of the collection funnel. This is also effective in blocking any errant feces pellets from the urine vessel. The glass wool must, however, be checked for adherence of the chemical being tracked.

D. CLOSED SYSTEM CAGES

Closed system cages, made of pyrex glass or plastic, are also available (e.g., Harvard) for collecting respiratory gases from rodents, in addition to excreta. After the animal has been treated by parenteral injection it is placed in the metabolism cage. The system is sealed and conditioned air is drawn through the top of the cage and exits through a port at the bottom. The air flow may be 200 to 500 ml/min depending upon the size animal and cage. The exit gas is then drawn through collection tubes containing a sorbant material. In cases where a ^{14}C-labeled test substance is used, CO_2 is the target gas and is trapped in scrubbers containing either ethanolamine, 0.5% NaOH, or a commercial sorbant (Oxifluor™-CO_2; DuPont-NEN, Boston, MA). Many investigators use the expired gases system in screening tests to assure molecular stability of the ^{14}C label prior to skin absorption studies. If the label is easily removed as $^{14}CO_2$, the site of label should be altered. A Tenax trap, placed just ahead of the CO_2 collector, has been used to differentiate between expired volatile test substance and that metabolically converted to $^{14}CO_2$.[15] This technique may be applicable to topical applications with volatile materials; thus providing a distinction between that fraction arising from the skin surface (unabsorbed) and that evolving from the lungs (absorbed). Preliminary tests would be needed to first identify the origin of volatile material (skin or lungs) having an affinity for the Tenax absorber.

Carbon dioxide may be collected from animals too large for the standard closed system metabolism cages. One reported study describes a gas mask for rabbits which fits over the

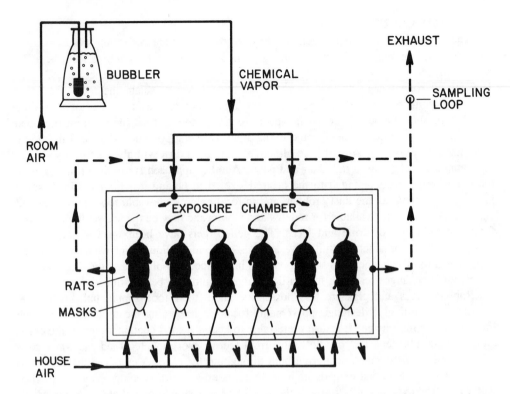

FIGURE 4.   Exposure system for rats when measuring percutaneous absorption of airborne materials.

animal's mouth.[16] Expired air was drawn from the mask through two serially connected traps containing 1 N NaOH. Animals were anesthetized during the 15- to 30-min exposure. Another study used anesthetized dogs in determining evolved $^{14}CO_2$ following an intravenous injection of radiochemical.[17] Dogs were intubated with an endotracheal tube fitted with a 2-way valve. Following treatment, exhaled air was collected in 20 l plastic bags attached to the 2-way valve. When filled, each bag was evacuated through a series of scrubbers containing a CO_2 absorber/scintillant (e.g., Oxifluor-CO_2).

VI. SPECIAL APPLICATIONS

Occasionally it becomes necessary to expose animals to dermal penetrants in a manner closely simulating an occupational or therapeutic exposure. These studies are designed to answer specific questions posed by the product developer or by the regulatory community. They may also be necessary to delineate the sorptive action of a unique chemical substance or to simulate an unusual use pattern. Three different studies are described as examples.

A. AIRBORNE EXPOSURES

Airborne exposures to chemicals have been routinely performed in animals to assess toxicity in terms of inhaled dose. Of continuing concern is that part of the toxic insult which results from simultaneous skin exposure and absorption. This prompted the development of a method for measuring whole-body vapor exposure (skin absorption), independent of inhaled dose.[18] Rats were closely clipped 24 h before exposure. The next day they were fitted with a latex mask held in place by a harness just behind the front legs. The mask had two breathing ports close to the nose compartment such that room air, under positive pressure, could be passed through the mask. The rats were restrained inside a specially designed vapor exposure chamber (Figure 4). For this study, a jugular cannula was also implanted in each rat such that blood levels of the airborne vapor (or its metabolites) penetrating the skin could be

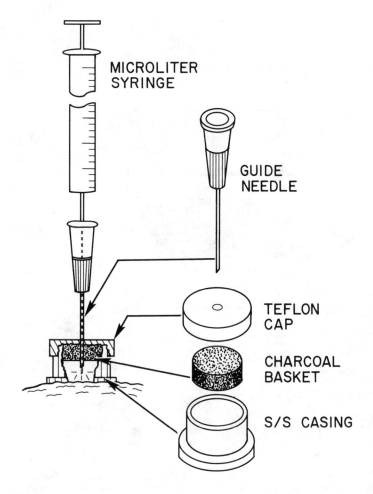

MICROLITER
SYRINGE

GUIDE
NEEDLE

TEFLON
CAP

CHARCOAL
BASKET

S/S CASING

FIGURE 5.   "Skin-Depot" for the collection of volatiles arising from topically applied substances.

measured. The animals were exposed for 4 h. This innovative study reported excellent continuity of results among the animals tested and demonstrated that the skin is a significant portal of entry for certain airborne chemicals.

B. VOLATILE SUBSTANCES

The evaluation of skin absorption of highly volatile substances is difficult at best. Often, these materials are only slightly absorbed due to their rapid vaporization. Skin penetration, therefore, occurs very early in the exposure period then declines rapidly as the concentration gradient is reversed. The classic excreta analyses suffer from poor accountability of the test substance. An improved method (Figure 5) uses a "skin-depot" to trap that fraction of material which would normally be lost by evaporation.[19] The depot is a short section of stainless steel tubing, 1 cm in diameter. Fitted into the top of the tube is a wire mesh basket containing about 100 mg of activated charcoal. A teflon cap covers the depot. A guide needle penetrates the cap and charcoal depot to allow skin application with a Hamilton (Hamilton Company, Reno, NV) syringe. The depot is attached to the animal's back with cyanoacrylate glue. After the test substance is applied, the guide needle and syringe are removed. The animal is immediately placed in a closed system metabolism cage for the collection of expired air and excreta. The study duration, usually a few hours, will be influenced by the volatility of the test substance and capacity of the depot. This methodology

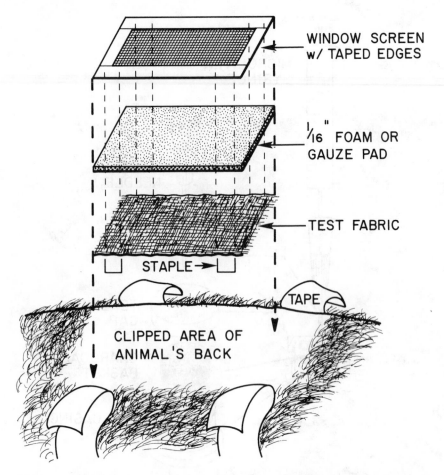

FIGURE 6.   Protective covering for swatches of clothing being assessed for transfer of fabric impregnants to the skin surface.

allows direct quantitation of absorbed chemical (urine, feces, and evolved CO_2), as well as evaporated chemical trapped in the charcoal sorbant. Accountability is further enhanced by measurements of material remaining on the skin surface and in the carcass at the end of the test, plus that recovered from the depot components.

C. FABRIC IMPREGNANTS

Fabric treatments are often used to extend the useful life of textiles, and in the case of clothing, to enhance the well-being of the wearer. Treatments may include fire retardants, sizing materials, fungicides, insecticides, water repellents, etc. Of concern is the potential for migration of clothing impregnants to the skin surface of the wearer and possible absorption. This issue was addressed using military uniform fabrics impregnated with a candidate insecticide/insect repellent.[20] Swatches of cloth, 50 cm² square, were treated with a solution containing ^{14}C-labeled permethrin. When dry, each swatch was covered with a thin (1/16 in.) piece of nonocclusive foam sheeting (similar to the filter materials used in window air conditioners), then an outer layer of wire window screen (Figure 6). The foam sheeting prevented any surface radioactivity from escaping the cloth but did not impede evaporation. The screen discouraged probing or chewing by the rabbit. The fabric was stapled at the corners through the foam sheet and wire screen. The pad-like device was then affixed, fabric side down, to the backs of rabbits using adhesive tape around the trunk and along the sides. Rabbits wore the cloth swatches continuously for 1 or 3 weeks. Impregnant loss from the fabric was measured as ^{14}C label appearing in excreta plus that residing on the skin surface at the end of the test period. The test substance remaining in the fabric following 1 to 3 weeks of wear was also measured. A similar approach may be used to examine the effects of accidental chemical spills on clothing (an everyday industrial occurrence) and the resulting percutaneous absorption. For this model, screen-covered fabric is affixed to the animal, then a metered liquid dose applied through the screen. No intermediate layer of foam sheeting or gauze is used.

VII. DISCUSSION

The preceding commentary on animal handling, protective devices and metabolism cages represents the collective experiences of many investigators — more or less a consensus. It is not intended to inhibit technical creativity for there is much to be done. Even in failure some unexpected insight is often gained. For example, we recently tried using sections from the legs of pantyhose as a body sleeve for rabbits. The intent was to secure swatches of impregnated fabric to the animal's back under a nonocclusive covering which would be unrestrictive and well tolerated by the animal for periods up to 3 weeks. Each section of nylon sleeve was about 8 in. long and open at both ends. Four leg holes were cut into each one so that when worn by the rabbit, the sleeve would remain relatively in place. A swatch of fabric impregnated with ^{14}C-permethrin was placed between the sleeve and the shaved back of the rabbit and was stapled outwards through the test fabric and the nylon covering. About half the rabbits wore the ensemble for two weeks with little apparent interest, but the remaining animals shredded the nylon sleeve within 48 h. The dyes used in the nylon were also a problem. These were readily extractable in methanol which was being used to remove residual radioactivity from the sleeves at the end of the test. While the colors were striking (the pantyhose had been donated during the holiday season), specimens were useless for scintillation counting. The one significant study result was that percutaneous absorption of the test fabric impregnant was increased nearly tenfold when used under the nylon sleeve, compared to results from studies using the standard gauze/screen covering. The increased absorption may have been caused by elevated temperatures (a known effect of pantyhose wear), or possibly by the abrasive action of the test fabric held snuggley to the skin surface for 2 weeks.

The true test of a skin absorption study in animals, or degree of percision, rests upon consistency of the data and accountability. Because of the small number of animals used, the inherent variance in recoveries from excretion, extractions, etc., must be minimal. This is not an unreasonable goal. Accountability of the applied test material (mass balance) should approach 100%. This is particularly true when rodents are used since their physical size is not a marked dilution factor; the entire carcass can be analyzed. Tests where recoveries of applied substance are less than 90% need to be technically reviewed. That is not to say that lower recoveries are unacceptable, for there have times in our laboratory when 80% would have been cause for celebration. However, the prudent investigator should strive for perfection and continually refine his methods to that end.

Finally, it is important that the results of the animal study be kept in perspective; that the data be realistically presented and analogies tempered. If skin were a simple membrane, absorption kinetics in both man and animals would, in all likelihood, be identical. However, they are not, so exploit the similaries and acknowledge the differences.

VIII. SOURCE LIST

Glass Metabolism Cages (rodent)

Allentown Caging Equip. Co., Inc.
P.O. Box 698, Route 526
Allentown, NJ 08501-0698

Harvard Apparatus, Inc.
22 Pleasant Street
South Natick, MA 01760

Bio-Serve, Inc.
P.O. Box 450
Railroad Avenue
Frenchtown, NJ 08825

Braintree Scientific, Inc.
P.O. Box 361
60 Columbia Street
Braintree, MA 02184

Plastic Metabolism Cages (rodent)

Maryland Plastics Inc.
251 E. Central Avenue
Federalsburg, MD 21632

Nalge Company
P.O. Box 20365
Rochester, NY 14602-0365

IITC Inc./Life Science
23924 Victory Blvd.
Woodland Hills, CA 91367

Plas-Labs, Inc.
917 E. Chilson
Lansing, MI 48906

Metabolism Cages (nonhuman primate)

Bush Products
1520 Cavitt Street
Bryan, TX 77801

Columbus Instruments
950 North Hague
P.O. Box 44049
Columbus, OH 43204

Primate Products
16230 Skyline Blvd.
Woodside, CA 94062

Suburban Surgical Co., Inc.
275 Twelfth Street
Wheeling, IL 60090

Custom Animal Diets

Bio-Serv, Inc.
P.O. Box 450
Railroad Avenue
Frenchtown, NJ 08825

Teklad Diets
P.O. Box 4220
2826 Lathham Drive
Madison, WI 53713

United States Biochemical Corp.
P.O. Box 22400
Cleveland, OH 44122

REFERENCES

1. **Gummer, C. L. and Maibach, H. I.,** Animal models for percutaneous penetration: assessing contamination of the experimental system, *Br. J. Derm.,* 115, 335, 1985.
2. **Bartek, M. J., LaBudde, J. A., and Maibach, H. I.,** Skin permeability *in vivo:* comparison in rat, rabbit, pig and man, *J. Invest. Dermatol.,* 58, 114, 1972.
3. **Pang, V. F., Swanson, S. P., Beasley, V. R., Buck, W. B., and Haschek, W. M.,** The toxicity of T-2 toxin in swine following topical application, *Fund. Appl. Toxicol.,* 9, 41, 1987.
4. Unpublished technique used by the author.
5. **Reifenrath, W. G., Chellquist, E. M., Shipwash, E. A., Jederberg, W. W., and Krueger, G. G.,** Percutaneous penetration in the hairless dog, weanling pig and grafted athymic nude mouse: evaluation of models for predicting skin penetration in man, *Br. J. Dermatol.,* 111, Suppl. 27, 123, 1984.

6. **Aldrich, F. D., Busby, W. F., Jr., and Fox, J. G.,** Excretion of radioactivity from rats and rabbits following cutaneous application of two ¹⁴C-labeled azo dyes, *J. Toxicol. Environ. Health,* 18, 347, 1986.

7. **Snodgrass, H. L., Nelson, D. C., and Weeks, M. H.,** Dermal penetration and potential for placental transfer of the insect repellent, *N,N*-diethyl-*m* toluamide, *Am. Ind. Hyg. Assoc. J.,* 43(10), 747, 1982.

8. **Shah, P. V., Fisher, H. L., Sumler, M. R., Monroe, R. J., Chernoff, N., and Hall, L. L.,** Comparison of the penetration of 14 pesticides through the skin of young and adult rats, *J. Toxicol. Environ. Health,* 21, 353, 1987.

9. **Scott, R. C. and Ramsey, J. D.,** Comparison of the *in vivo* and *in vitro* percutaneous absorption of a lipophilic molecule (cypermethrin, a pyrethroid insecticide), *J. Invest. Dermatol.,* 89, 142, 1987.

10. **Bronaugh, R. L., Stewart, R. F., Congdon, E. R., and Giles, A. L., Jr.,** Methods for *in vitro* percutaneous absorption studies. I. Comparison with *in vivo* results, *Toxicol. Appl. Pharmacol.,* 62, 474, 1982.

11. **Simoneaux, B.,** Skin Penetration of Pesticides, presented at Ninth Annual Meeting of the American College of Toxicology, Baltimore, Oct. 31 — Nov. 2, 1988.

12. Personal communication with Mr. Sidney Sprague, Loctite Corp., Newington, CT.

13. **Derelanko, M. J., Gad, S. C., Gavigan, F. A., Babich, P. C., and Rinehart, W. E.,** Toxicity of hydroxylamine sulfate following dermal exposure: variability with exposure method and species, *Fund. Appl. Toxicol.,* 8, 583, 1987.

14. **Shah, P. V., Sumler, M. R., Ioannou, Y. M., Fisher, H. L., and Hall, L. L.,** Dermal absorption and disposition of 1,3-diphenylguanidine in rats, *J. Toxicol. Environ. Health,* 15, 623, 1985.

15. **Klain, G. J., Bonner, S. J., and Omaye, S. T.,** Skin penetration and tissue distribution of [¹⁴C]butyl 2-chloroethyl sulfide in the rat, *J. Toxicol. Cut. Toxicol.,* 7(4), 225, 1988.

16. **Hashimoto, K. and Tanii, H.,** Percutaneous absorption of [¹⁴C] methacrylamide in animals, *Arch. Toxicol.,* 57, 94, 1985.

17. **Snodgrass, H. L. and Nelson, D. C.,** Dermal penetration of ¹⁴C-labeled permethrin isomers, Study No. 75-51-0351-83, *U.S. Army Environmental Hygiene Agency Report,* Aberdeen Proving Ground, MD, July 15, 1983.

18. **McDougal, J. N., Jepson, G. W., Clewell, H. J., III, and Anderson, M. E.,** Dermal absorption of dihalomethane vapor, *Toxicol. Appl. Pharmacol.,* 79, 150, 1985.

19. **Susten, A. S., Dames, B. L., Burg, J. R., and Niemeier, R. W.,** Percutaneous penetration of benzene in hairless mice: an estimate of dermal absorption during tire-building operations, *Am. J. Indust. Med.,* 7, 323, 1985.

20. **Snodgrass, H. L. and McGreal, P. A.,** Migration of permethrin from military fabrics under varying environmental conditions, Study No. 75-52-0687-88, *U.S. Army Environmental Hygiene Agency Report,* Aberdeen Proving Ground, MD, Sept. 13, 1988.

Chapter 8

DETECTION METHODS

Marie Lodén

TABLE OF CONTENTS

INTRODUCTION

Substances penetrating the skin can be quantified using three different methods. First, radiometric assays measure the radioactivity, irregardless of its association with the parent molecule, metabolite, or radiochemical impurity. Second, chemical assays quantify specific substances. Third, biological assays measure the pharmacological activity of the applied compound together with active metabolites.

The radiometric assay is by far the most simple and widely used method to analyze penetrants. The technique facilitates analysis of chemicals in body tissues, as well as receptor fluids from permeability studies in vitro. It is also possible to localize penetrants in different skin structures by autoradiography.

It is necessary to be aware of the limitations and possible pitfalls of radiometric assays. The analysis cannot distinguish between parent compound and its decomposition products. Not all compounds are readily available as isotopes. Ethical considerations must be made when the experiments are done with humans. Radiochemical detection is most appropriate when isolated human skin is used and when the experiments are performed on animals. The first section of the chapter will focus on the liquid scintillation technique to measure skin permeability.

Chemical methods are not as simple as radiometric methods. Usually the components in the sample must be separated before quantification. There are a variety of separation systems and detectors to choose between, and it may not be obvious to the investigator which combination to use for a certain substance. However, most substances can be analyzed by several methods and the primary considerations are probably the equipment available and the personal preference. Generally it is of value to combine different methods to increase sensitivity and specificity of the analysis. In the second section of this chapter the reader will be introduced to common separation methods of gas, high-pressure liquid, and thin-layer chromatography (GC, HPLC, and TLC, respectively) and detectors.

Biological assays may be appropriate for substances which primarily exert their effects in the skin (steroids, surfactants, etc.). For some substances a biological assay is the only suitable alternative to chemical and radiometric analyses, due to high specificity and sensitivity. In the third section a few examples of biological assays will be given.

II. RADIOCHEMICAL DETECTION

A. GENERAL ASPECTS
Radioactive isotopes are used by many investigators to study the permeability of the

TABLE 1
Properties of Some Beta Emitting Radionuclides

Radionuclide	Half-life	Maximum beta energy (MeV)	Maximum specific activity
Tritium	12.4 years	0.0186	1.07×10^{12} Bq/mA[a]
Carbon-14	5730 years	0.156	2.31×10^9 Bq/mA
Sulfur-35	87.4 d	0.167	55.3×10^{12} Bq/mA
Phosphorus-32	14.3 d	1.709	338×10^{12} Bq/mA

[a]   mA = milligramatom.

skin. Isotopes act as tracers for chemically similar nonradioactive compounds. Depending on the energy of the isotopes and their type (alpha-, beta-, and/or gamma ray emitters) different detector system are used. Gas ionization counters, semiconductor detectors, Geiger-Müller counters, and solid scintillation counters are used to measure isotopes which emit gamma rays and/or medium to high energy beta particles. Liquid scintillation counting is the most efficient method to measure low energy beta emitters, such as carbon-14, tritium (hydrogen-3), and sulfur-35. Compounds labeled with beta emitters are the most commonly used isotopes in skin permeability experiments. Some properties of beta emitters are listed in Table 1. Several textbooks have been written on the fundamentals of radiochemistry, radiation detectors, and applications of isotope methods of analysis.[1-4] The main suppliers of liquid scintillation counters and isotopes are listed in the Appendix. In the following section the liquid scintillation technique will be described.

B. DEFINITIONS
A scintillator converts energy from nuclear decay into light, which is analyzed and displayed as counts or as count rate. The count rate is a liquid scintillation counter is usually expressed in counts per minute (CPM) which can be converted into absolute units of disintegrations per minute or DPM. One nuclear disintegration per second is equal to one Becquerel (Bq), according to the units of radioactivity in the International System of Units (SI). Another frequently used unit is Curie (Ci), and one Curie equals 3.7×10^{10} Bq.

Useful conversion factors are

$$1 \text{ DPM} \qquad = 60 \text{ Bq} \qquad = 16.2 \times 10^{-10} \text{ Ci} \qquad (1)$$

$$16.7 \times 10^{-3} \text{ DPM} = 1 \text{ Bq} \qquad = 0.27 \times 10^{-10} \text{ Ci} \qquad (2)$$

$$617 \times 10^6 \text{ DPM} = 3.7 \times 10^{10} \text{ Bq} = 1 \text{ Ci} \qquad (3)$$

The specific activity is defined as the activity per unit mass of an element or compound containing a nuclide, i.e., mCi/mmol or mCi/mg. Radioactive concentration relates to the amount of active compound dissolved in a solvent, i.e., mCi/ml.

C. COUNTING EFFICIENCY
The efficiency in converting the incident radiation energy of a sample to the detected count rate is determined by the quenching in the sample. The quenching can be of two kinds: color quenching, which is due to absorbance of the emitted light, and chemical quenching, which reduces the energy transfer between the solvent and the scintillation molecules.

If the counting efficiency is known, then the radioactive content can be expressed in absolute units (DPM) by simply dividing the observed count rate (CPM) by the counting efficiency. The counting efficiency can be determined from a quench correction curve or by using an internal standard method. All commercial counters today contain microprocessors which automatically convert the sample count into the estimated absolute activity of the sample, including appropriate error limits for the reported value.

1. Quench Correction Curve

To establish a quench correction curve an unquenched standard with known activity of the isotope is required. Calibrated standards (such as ^{3}H hexadecane and ^{14}C hexadecane) and scintillation cocktails are obtained from commercial sources (see Section II.D). Scintillation cocktail is dispensed into approximately ten counting vials. To each of the vials an exact amount of a calibrated standard is added and the CPM is measured. Variation in count rates among the vials represent errors in pipetting the standards, or contamination. An increasing amount of quencher (e.g., carbon tetrachloride, acetone) is then added to each vial. From the loss in count rate after the addition of the quencher, the counting efficiency of each vial can be calculated and plotted against a quench indicating parameter. Such a series of quenched standards with known activity may also be obtained commercially.

2. Internal Standard Method

The counting efficiency can also be determined by an internal standard method, which involves addition of a nonquenching radioactive standard to a sample that has been previously determined for CPM. From the addition CPM, the efficiency of the sample can be calculated. The DPM of the actual sample is obtained by division of the CPM by the efficiency. The advantage of the method is that it is simple and long established. The disadvantage is the required manipulation of the samples which are time consuming for long series.

D. PREPARATION OF SAMPLES

Aqueous samples (e.g., receptor fluids) can often be dissolved directly in an emulsifier-type scintillation cocktail. There are many cocktails available which can incorporate large amounts of water. The sample should be homogeneous during counting (must be checked after refrigeration of the sample in the counter). Deciding which cocktail to use is usually a compromise between counting performance (sample holding capacity and counting efficiency) and cost. Generally, the lower cost cocktails produce lower counting performance. Another important concern is the safety of the cocktail. A scintillation cocktail which has a low vapor pressure and a flash point above the laboratory temperature, is less of a health and fire hazard than a cocktail with a high vapor pressure and a low flash point. Among the cocktails which are considered environmentally safe and have high counting efficiency for aqueous samples are Ecolume (ICI), Ready Safe (Beckman), OptiPhase Safe (Pharmacia Wallac), and Ultima-Gold (Packard).

Most biological samples (body tissues, feces, blood, etc) are not readily soluble in the commonly used solvents of the scintillation cocktails. Special techniques, such as solubilization and oxidation, are required to obtain a homogeneous system for reproducible measurement of the radioactive sample. The tissue can be solubilized by hydrolysis with NaOH, KOH, formamide, or quaternary ammonium bases. Of these, the quaternary bases have strong solubilizing power and are widely used in commercial solubilizers, e.g., Soluene-100 and -350 (Packard), BTS 450 (Beckman), OptiSolv (Pharmacia Wallac), Protosol (du Pont NEN), and NCS (Amersham).

Solubilizing may cause color quenching and chemiluminiscence. If decolorization is considered necessary, the sample can be bleached with hydrogen peroxide (200 μl of 30 to 35% hydrogen peroxide for every 100 g wet tissue). Chemiluminiscence is the production of light as a result of a chemical reaction. This most typically occurs in samples of alkaline pH and those containing peroxides. Neutralisation of alkaline samples with, e.g., acetic or hydrochloric acid can dramatically reduce the level of chemiluminiscence. Reduction of chemiluminiscence in peroxide containing samples can be achieved by adding isopropanol to a final concentration of 3 to 5% in the scintillation cocktail.

Sample oxidation is recommended when solubilized material are intensely colored or when the radioactive concentration is very low. The oxidation of samples containing ^{14}C, ^3H, and ^{35}S will liberate $^{14}CO_2$, ^3H_2O, and $^{35}SO_3$, respectively, which are collected and incorporated into scintillation cocktails for counting. An automated combustion unit is marked by Packard as Tri-Carb 306 and by Harvey Instrument as Harvey Ox-300.

E. PURITY AND STABILITY OF ISOTOPES
1. Purity

Chemical purity of a material may be defined as the percentage of the mass of material present in the specified chemical form. Radiochemical purity is the proportion of the isotope in a stated chemical form. Thus, impurities may or may not be radioactive. The fraction of molecules in a bulk of material which is labeled with a radionuclide is usually very small (<1%).[1] The amount can be calculated from the maximum specific activity of the isotope (Table 1) and the specific activity of the compound. Thus, even if all radiolabelled molecules are in the wrong chemical form, this could not contribute to any significant extent to the chemical purity.

2. Stability

The quality of a radiolabeled compound is normally highest immediately following its preparation and purification but because of self decomposition this quality is not maintained on storage. The presence of radioactive atoms in a molecule frequently makes the compound more unstable compared to the nonlabeled molecule.

Radiation decomposition is related to the specific activity and the radioactive concentration.[5] In general, at high specific activity, the rate of decomposition is proportional to radioactive concentration and independent of specific activity, while at low specific activity it is proportional to specific activity and less dependent on radioactive concentration. Decomposition can be minimized by diluting the radioactive molecules with unlabeled molecules and by reducing the number of solute-radical interactions. The latter can be achieved by lowering the temperature of storage and by the addition of radical scavengers (e.g., ethanol, benzyl alcohol, glycerol).

Freezing of a solution may cause molecular clustering of the solute if the pure solvent around the edge of the sample freezes first. This effect has a great impact on the decomposition of tritium labeled compounds.[1] In general, tritium labeled compounds are much more unstable than compounds labeled with radionuclides emitting higher radiation energy.[1] This is due to the usually high specific activity of tritium compounds and the fact that virtually all of the decay energy is deposited within the vicinity of the tracer. Hence the stability of tritiated compounds in solutions is usually higher when stored just above 0°C than at −20°C.

F. EFFECTS OF IMPURITIES IN PERMEABILITY STUDIES

When using isotopes in skin permeability experiments the investigator should be aware of the problems associated with radiolabeled impurities in the compound. Absolute purity cannot be achieved in practice. What is essential is that the impurities do not affect the interpretation of the experimental data. The purity of most substances can easily be checked by thin-layer chromatography (see Section III.E.2). The supplier of an isotope can often recommend suitable systems for checking the purity.

It is seldom possible to predict the effects of radiolabeled impurities on the results obtained from experiments using impure radiochemicals. However, if the composition of the impurity is known, then a rough estimate regarding its influence on the result can be made. The estimate is based on the known difference in permeability of the skin to the penetrant and to the impurity.

Marzulli et al. proposed a scheme for rating substances as to their capacity for penetrating human skin according to their permeability constant (Kp value).[6] Five categories were provided, covering five orders of magnitude, with descriptive terms assigned to each category. For example, the Kp of a very slow penetrating substance is <0.1 and of a moderate 1 to 10 μcm min^{-1}. The lower the Kp of the penetrant, the higher radiochemical purity is required to obtain reliable results. If the Kp of the impurity is lower than the Kp of the penetrant, the maximum error will be as large as the percentage impurity in the applied compound; that is, a maximum 10% lower penetration rate when 10% radiolabeled impurity is present. An overestimation of the penetration rate is obtained if Kp of the impurity is higher than that of the penetrant. A very slow penetrating substance may well contain impurities with much higher penetrating capacity. For example, if the sample contains 10% of a radiolabeled impurity with a Kp 100 times higher than the Kp of the penetrant of interest, then the resulting Kp will be about 10 times higher than the Kp of the pure agent. Hence, one must be careful in evaluating the result when slowly penetrating substances are studied.

When permeability studies are performed *in vitro* there are two practical methods to elucidate the influence of impurities on the result. One method is to isolate the impurities and to actually test their penetration properties. The other method is to analyze chemically the amount of penetrant in the receptor fluid and compare the obtained result with the result from the radiochemical detection. A good agreement between the results implicates absence of interfering impurities. A discrepancy between the result from the two detection methods may either reflect radiolabeled impurities in the applied compound, or the fact that the penetrant was decomposed in the skin during penetration.[6-8]

The influence of radiolabeled impurities can be exemplified by a study of heparin.[9] Heparin is a highly sulfated polysaccharide with a molecular weight of about 15,000 Da. Hence, the substance was expected to penetrate the skin with extreme difficulty. Permeability experiments with tritium labeled heparin gave a Kp of approximately 1 μcm min^{-1}. The radiochemical purity was 95%. The radiolabeled impurities were isolated by gel filtration and applied to the skin. The impurities were found to have a Kp of about 20 μcm min^{-1}. Thus, the measured Kp of heparin was erroneously due to the much higher penetration rate of the impurities than of heparin.

II. CHEMICAL DETECTION

Chromatography is essentially a physical method of separation in which the components to be separated are distributed between two phases, a stationary bed (a solid or a liquid) and a fluid or a gas which moves through or along the stationary bed. There are two major classes of chromatography depending on the mobile phase: liquid chromatography (LC) and gas chromatography (GC).

There are several textbooks and journals devoted exclusively to chromatography (*Journal of Gas Chromatography, Journal of Chromatography, Journal of Chromatographic Science, Chromatographia, Journal of High Resolution Chromatography & Chromatographic Communications, Biomedical Chromatography,* and *Journal of Liquid Chromatography*).[10-15] The bibliography section in the *Journal of Chromatography* provides the investigator with a continuing review of the chromatography literature. Commercial suppliers of chromatographs and accessories can be found in the *LabGuide of Analytical Chemistry* and in the chromatography guide issue of the *Journal of Chromatographic Science*.

A. PREPARATION OF SAMPLES

Before measuring a certain substance in a complex sample (body tissue, blood, urine, etc.) constituents that might damage the analytical equipment or interfere with detection must be removed. Methods for sample preparation and detection have been described in many textbooks and reviews.[16-22]

Tissues must be homogenized before a sample is taken. The material can be reduced by grinding under liquid nitrogen, by slicing in a freeze-microtome or by homogenization with a homogenizer. The homogenate plus solvent can be centrifuged and the supernatant sampled for analysis.

Usually the proteins in an aqueous sample must be removed prior to analysis. This can be done by precipitation of the proteins by addition of strong acids (e.g., hydrochloric acid, trichloroacetic acid, or perchloric acid) followed by centrifugation.

1. Liquid-Liquid Extraction

The conventional technique to purify samples is liquid-liquid extraction. The crude homogenate, its supernatant, blood plasma, the deproteinized sample, etc., can be processed by this method.[20] Liquid-liquid extraction is most useful for substances of lipophilic character extracted in the uncharged form from an aqueous phase. Maximum extraction is generally achieved with solvents whose polarity is similar to the polarity of the compounds being extracted. For a compound with hydrogen accepting properties, a hydrogen-donating solvent is preferred, and vice versa. When the organic solvent is chosen carefully the technique can be very selective. In the case of organic ions, extracting into the organic phase can be facilitated by ion pairing. By varying the counterion, the degree of extraction can be varied within wide limits. The recovery usually increases with increasing hydrophobicity of the counterion.

Adjustment of the pH of the sample is used to increase the degree of extraction of acids and bases. Weak acidic substances are usually extracted at a pH of about 6, whereas for stronger acids pH must be adjusted to about 2 for maximum recovery. For weak bases, extractions are usually carried out at a pH of 7 to 8, for some amphoteric agents at 8.5 to 9, and at pH 10 or higher for strong basic substances. The association of some drug substances with proteins can decrease the extraction efficiency. The recovery can often be increased by dilution of the sample with buffer solutions or by degradation of the proteins by addition of certain enzymes.

After shaking two immiscible liquids emulsions can be formed. The tendency to form emulsions increases with liquids of relatively high mutual solubility, and with high-viscosity solvents and solvents with densities similar to that of water. Emulsions may be broken by adding neutral salts, refrigeration, or centrifugation. Following extraction of the sample, the solvent phase may have to be further treated prior to analysis (e.g., filtration, drying over anhydrous sodium sulfate, or concentration by evaporation).

2. Liquid-Solid Extraction

Another way to purify compounds in biological samples is to use liquid-solid distribution. Typical solid phases are silica, bonded-silica, alumina, polystyrene-based polymers, and dextran-based gels. Several manufacturers supply prepacked cartridges of these materials in a form that can be used with a syringe for charging sample and solvent to the cartridge. A typical purification involves passage of the sample through the cartridge (allows retention of the compound), removal of undesirable matrix components by washing with buffer solutions and then elution of the sample by desorption with a suitable solvent. Before injection onto a column, it may be necessary to concentrate the extract by evaporation.

B. STABILITY AND RECOVERY OF COMPOUNDS

The stability of the substance during storage and sample preparation must always be checked. The decomposition of compounds is due to enzymatic or chemical reactions. Decomposition due to enzymes may be avoided by a change of pH of the sample, by addition of enzyme inhibitors, or by denaturation of proteins. Chemical decomposition, such as hydrolysis or oxidation, can be controlled by adjustment of pH, by addition of antioxidants and by lowering of the storage temperature. The effects of repeated freezing and thawing must also be considered.

Many compounds can be lost during the preparation due to adsorption to different surfaces. The risk seems to be most pronounced for hydrophobic compounds which contain a polar function like amine, phenol, or alcohol.

The recovery can be checked by adding a known amount of substance to the unprocessed biological sample, and comparing the results from direct injection of pure solutions of substance with that from the processed biological sample.

C. DERIVATIZATION

Derivatization is frequently used in GC since many compound cannot be chromatographed directly, due to a lack of volatility or thermal stability. In both GC and LC, derivates are prepared to achieve the desired sensitivity or specificity of a given separation. Ideally, derivatization should be rapid and quantitative with a minimum of by-products. Excess reagents should not interfere or should be easily removed.

Derivatization can be carried out either prior to or after the separation (pre- or post-column).[17,23] This can be done continuously as a part of a chromatographic system, or in a batch procedure, separate from the chromatographic step.

D. STANDARDS, CALIBRATION

Quantitative analysis involves conversion of the detector signal to a known quantity. Standard addition, external standard, and internal standard are the techniques usually selected.

1. Standard Addition

The method of standard addition can be useful when an analyte free sample cannot be obtained for constructing a calibration curve, or when the loss of the compound during sample preparation is unknown. The sample is first submitted to analysis and the response measured. A known amount of the compound to be determined is then added to the sample, and a further analysis is made. The ratio between the responses is used to calculate the amount of substance in the original sample.

2. External Standard

A calibration curve is constructed when the external standard method is used. The curve is obtained by injecting known amounts of the compound and plotting the response against the injected quantity. The sample is then injected and from the response the amount of the substance can be read from the calibration curve. Disadvantages of this method are that the precise amount of sample injected must be known and that calibration is time consuming. Also the sensitivity of the detector must remain constant from run to run.

3. Internal Standard

A widely used technique of quantification involves the addition of an internal standard to the sample to compensate for analytical errors, such as detector drift and fluctuations in injection volumes. The compound chosen as internal standard should have structural similarity to the studied compound, similar extraction efficiency and elute close to its peak.

Known weight ratios of the compound and the internal standard are prepared and chromatographed. The responses are measured and their ratios are plotted against the respective

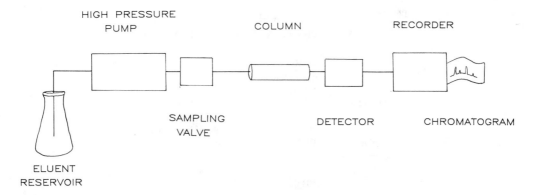

FIGURE 1. General schematic of an HPLC system.

weight ratios to obtain a graph. An accurately known amount of the internal standard is then added to the sample and this mixture is chromatographed. Response ratios are measured, and from the calibration graph the weight ratio of the studied compound to the standard is read. Since the amount of standard added is known, it is a simple calculation to determine the amount of the studied compound. The advantages of internal standardization are that the sample quantities chromatographed need not be accurately measured, and that the detector response need not be known or remain constant, since any change in response will not alter the response ratio. The chief disadvantage of the method is the difficulty in finding a standard that does not interfere with the unknown compound in the sample.

E. SEPARATION OF COMPOUNDS
1. Liquid Chromatography

There are four basic liquid chromatography (LC) methods: liquid-liquid, liquid-solid, ion-exchange, and size-exclusion (gel filtration) chromatography. Modern LC techniques are usually called high-performance or high-pressure liquid chromatography (HPLC). HPLC is characterized by short separation times and offers particular advantages over GC in the separation of nonvolatile, thermolabile and very polar compounds. Other advantages include the possibility of full automation of the analysis and the fact that a variety of detectors are available. The technique and its applications have been described in many textbooks.[18,24-31]

A simple HPLC system requires a solvent reservoir (usually a reagent bottle), a high pressure pump, a sample injection valve, a column, a detector, and recording equipment (Figure 1).

Liquid-liquid or partition chromatography (LLC) involves a liquid stationary phase whose composition is different from that of the moving liquid phase. Sample molecules distribute between the mobile and stationary liquid phases, just as in liquid-liquid extraction within a separatory funnel. The moving- and stationary-phase liquids must be immiscible.

Bonded-phase chromatography, BPC, is a modification of liquid-liquid chromatography. The stationary phase is chemically bonded to particles instead of mechanically held as in LLC. The main advantage of BPC packings is that they are quite stable compared to LLC. As with LLC, the BPC separations is generally dependent on the type of functional groups within the solute molecule. Samples of moderate to strong polarity are usually well separated on the polar BPC packing. Reverse-phase BPC normally involves a relatively nonpolar stationary phase in combination with a very polar (e.g., aqueous) mobile phase to separate a wide variety of less polar solutes. Usually this allows injection of aqueous samples without further treatment.

Liquid-solid or adsorption chromatography involves high-surface-area particles, and retention of sample molecules by attraction to the surface of the particle. Thin-layer chromatography (TLC) is a classical form of liquid-solid chromatography (see Section III.E.2).

TABLE 2
Characteristics of LC Detectors

Detector	Minimum detectable concentration (g/ml)	Selectivity
UV	10^{-10}	UV-absorbing compounds
Refractive index	10^{-7}	Nonselective
Fluorometer	10^{-11}	Fluorescent compounds
Electrochemical	10^{-11}	Oxidizable compounds
Radioactivity	10^{-11}	Radionuclides

In ion-exchange chromatography the stationary phase contains fixed ionic groups along with counterions of opposite charge. The technique is widely used for separation of hydrophilic protolytes (acids and alkalis).

In size-exclusion chromatography (gel filtration) the column contains a porous material with pores of a certain size. Molecules that are too large are excluded from all the pores (elute early), whereas small molecules can penetrate most of the pores. Usually, separation is determined strictly by molecular size and it is the preferred method for separation of higher molecular weight substances (molecular weight >2000 Da).

a. Detectors

Several textbooks and reviews have been published on the principles and applications of LC detectors.[18,32-34] Some properties of commonly used LC detectors are listed in Table 2. The detectors most widely used in LC are based on absorption of ultraviolet (UV) and visible light.[35,36] Sample concentration in a flow cell is related to the fraction of light transmitted through the cell by Beer's law.

$$\log_{10} I_o/I = \epsilon bc \qquad (4)$$

where I_o is incident light intensity, I is the intensity of the transmitted light, ϵ is the molar absorptivity (or molar extinction coefficient) of the sample, b is the path length in cm, and c is the concentration in mol/l.

The UV-detectors are reliable and easy to operate. They are sensitive to UV absorbing substances, i.e., those having conjugated systems. The UV spectrophotometers allow operation at the absorption maximum of the compound, or at a wavelength that provide maximum selectivity. They have good linearity range (10^5). Potential disadvantages are the widely varying response to different compounds and that similar compounds may not be discriminated due to overlapping absorption.

The differential refractometer monitors the difference in refractive index (RI) between pure reference mobile phase and the column effluent. RI detectors respond to all compounds, are reliable and fairly easy to operate. However, such detectors have only limited sensitivity and are mostly used where sensitivity is not important.

The fluorometer is a very sensitive and selective detector for fluorescent compounds, excited by UV radiation. Compounds that contain aromatic rings or highly conjugated aliphatic systems often fluoresce. Nonfluorescent compounds can be detected after preparation of fluorescent derivatives (see Section III.C). In many cases the fluorescence detector is 100-fold more sensitive than UV absorption. Therefore, the fluorometer is one of the most sensitive LC detectors. The dynamic range of fluorometers can be fairly large (10^4) but the linear dynamic range may be less than two orders of magnitude. For quantitative work, it is strongly recommended that the detector be calibrated against known standards over the concentration range of interest.

Electrochemical (amperometric) detectors can selectively measure electroreducible and electrooxidizible compounds. Examples are oximes, hydroperoxides, aromatic amines, ketones, aldehydes, and nitro compounds. Compound types generally not sensed include ethers, aliphatic hydrocarbons, alcohols, and carboxylic acids. The great selectivity and sensitivity of the electrochemical detectors enhances its application for the analysis of known compounds in a complex matrix.

Radionuclides in the chromatographed sample can be detected and quantified using a radioactivity monitor. Solid scintillators and Cerenkov radiation can detect compounds labeled with high energy beta particles and gamma ray emitters, without losing too much counting efficiency (see Section II.C). To increase counting efficiency (especially important for low-energy beta sources) the eluate can be mixed with scintillation cocktail prior to detection. Mixing with cocktails also prevents memory effects, which can appear due to contamination of the solid scintillators. If it is unacceptable to mix the entire eluate with scintillation cocktail, then a splitter can direct a portion of the flow to another detector or to a fraction collector.

2. Thin-Layer Chromatography, TLC

In TLC the stationary phase is a finely divided solid, distributed as a thin layer on a glass or plastic plate. Most TLC separations are performed on the ''normal-phase'' mode on a polar stationary phase (e.g., silica gel, alumina, polyamide, or cellulose) and a nonpolar mobile phase. For separation of substances with high polarity, reversed-phase TLC is used. The technique employs a chemically bonded nonpolar stationary phase and a more polar mobile phase. Techniques and applications have been described in many textbooks.[13,15,37-39]

a. Development

The sample is applied to one end of the stationary phase. The plate is placed into a closed vessel containing a small amount of solvent which moves through the stationary phase by capillary action. Separation occurs when one substance in a mixture is more strongly absorbed by the stationary phase than the other components. Development is usually complete when the solvent front has travelled about three quarters of the length of the plate. The plate is then removed and dried.

2. Detection

Substances can be visualized by their fluorescence under long or short wavelength UV light, or by their ability to quench the fluorescence of an inorganic phosphor that can be added to the sorbent layer. The plate can also be treated with locating reagents to identify the separated compounds. Reagents for identification of substances can be found, for example, in the *CRC Handbook of Chromatography*.[40]

Radioactive substances can be located by placing the plate in contact with a sheet of X-ray film, which becomes blackened in the areas corresponding to the active spots. Radioactivity in the spots can be quantified by scraping the thin-layer material into vials and measuring the radioactivity. The most convenient way to quantitate the radioactivity is to use radioscanners which automatically determine the count-rate in the spots (see Appendix).

3. Gas Chromatography

In gas chromatography the stationary phase can be either a solid or a liquid. Gas-liquid chromatography, GLC, is more frequently used today. In GLC there are two kind of columns: capillary columns which are open tubes of a small diameter with a thin liquid film or a porous layer on the wall, and packed columns which consist of an inert solid material supporting a thin film of a nonvolatile liquid. Capillary columns have the high efficiency required for the separation of closely spaced peaks.

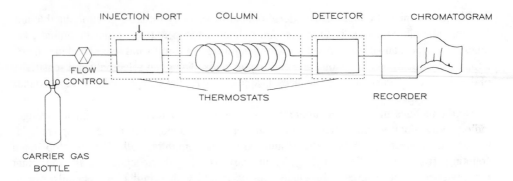

FIGURE 2.   General schematic of a GC system.

A gas chromatograph consists essentially of three integrated functions: sample introduction (evaporation), separation, and detection (Figure 2). General aspects on GC-methods and sample preparation can be found in several textbooks.[15,41-44]

A sample is introduced by a micro syringe (volume 1 to 25 μl) through a septum into the heated injector. The sample is then carried by the carrier gas into the column where the components are separated. The choice of column temperature usually improves resolution but increases the analysis time.

a. Sample Injection

When using packed columns the total amount of the injected sample is vaporized and transferred onto the column. In capillary chromatography, a variety of injection models have been developed, e.g., split, splitless, and on-column.[45,46] In split injection the major portion of the vaporized sample is discharged to the atmosphere via an outlet, while a minor portion is directed through the column. The reason for the split technique is to prevent flooding the column with solvent. The major disadvantages are that the sample is subjected to a relatively severe thermal shock, and that an incomplete transfer of the sample to the column can occur when the components are high boiling.

For samples of low concentration splitless injections may be used.[46] The sample is injected into the heated injector, usually at a slow rate. The vaporized sample is transferred into a cooler column where it condenses in the first part. An outlet to the atmosphere is then opened and the residual volatiles are vented from the injector. The sample is then successively eluted, leaving the substances in narrow bands. It may be possible to analyze thermally labile compounds since rapid vaporization of the sample is not required. The primary disadvantages are that the column efficiency can be rapidly lost, and that the portion of the sample directed into the column may not represent the true composition.

In on-column injection the sample is deposited as a liquid at the beginning of the open tubular column, at a temperature just below the solvent boiling point. This technique is favorable when working with thermolabile compounds and high boiling compounds, which may be decomposed or not transferred to the column in split and splitless injections, and for samples of low concentration.[45]

b. Detectors

All routinely used detectors are described in reviews and textbooks.[32,47] In Table 3 the characteristics of the most common GC detectors are given.

The two most important detectors in GC are the thermal conductivity detector and the flame ionization detector, FID. The thermal conductivity detector works on the principle that a hot body, a filament, will lose heat at a rate which depends on the composition of the surrounding gas. The change in temperature of the filament will give rise to a change

TABLE 3
Characteristics of GC Detectors

Detector	Minimum detectable quantity (g)	Selectivity	Linear range
Thermal conductivity	10^{-7}	Universal, except carrier gas	10^4
FID	10^{-10}	Organic compounds	10^6
Thermionic detector	10^{-10}	P, or N compounds	10^3
ECD	10^{-12}	Electronegative compounds	500

in electrical resistance. The detector is nondestructive and responds to any gas. The response depends on the concentration and on the molecular weight of the gas being detected.

FID is the most commonly used detector in GC. Components in the column effluent are ionized by burning in a hydrogen-air or oxygen flame. This allows the gas in the detector to conduct an electrical current. When a pure carrier gas (helium or nitrogen) is combusted, few ions are formed and the conductivity is low. The combustion of organic compounds in the carrier gas increases the conductivity and the resultant current is amplified and recorded. This detector is highly sensitive and its range of linear response to increasing concentration is very wide. The response is dependent on the carbon weight percentage in the molecule.

The thermionic detector (also known as the alkali flame ionization or nitrogen-phosphorus detector) is a modification of the FID by the addition of a small alkali salt pellet to the burner jet. This makes the detector sensitive to phosphorus or nitrogen compounds.

The most sensitive of the commonly used GC detectors is the electron capturing detector, ECD. Tritium and Ni^{63} are frequently used to ionize the molecules of the carrier gas such as nitrogen. These electrons migrate to the anode under a fixed voltage and produce a steady current. If a sample containing electron-absorbing substance is then introduced, this current will be reduced. The loss of current is a measure of the amount and electron affinity of the test compound. Thus, the electron capture is extremely sensitive to certain molecules such as alkyl halides, conjugated carbonyls, nitrates, nitriles, and organometallics but is virtually insensitive to others (e.g., hydrocarbons, alcohols, and ketones). The sensitivity for such compounds can be dramatically improved by chemical derivatization (see Section III.C).

The separated components at the end of the column may also be carried to a mass spectrometer (MS) for identification. In mass spectrometry the sample molecules are excited into ions, which are separated, characterized, and quantified. Despite the relatively high cost of GC-MS instrumentation, the combination of sensitivity and selectivity has led to a widespread adoption of GC-MS for measuring compounds in biological extracts. The technique is also used in combination with LC. Several textbooks have been published on the advantages and limitations of MS.[48-53]

c. Head-Space Analysis

Head-space analysis is used predominantly for the determination of volatile substances in samples which are difficult to analyze by conventional GC methods. The most important advantages with the head-space technique are the prevention of column contamination, the reduction of interference arising from the injection of large amounts of solvent, and enhanced sensitivity. The sample is placed in a vial of appropriate size and closed with a rubber septum. The temperature of the vial is carefully controlled until equilibrium between the sample and the gas phase is established. The gas phase is sampled by a syringe for manual procedures or with an electropneumatic dosing system in automated head-space analysis. The concentration of the compound in the gas phase may be increased by adjusting the pH

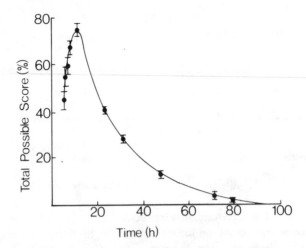

FIGURE 3.   Percentage total possible score as a function of time for Betnovate Cream:
mean of 10 experiments ± standard error of the mean. (From Barry, B. W. and Woodford,
R., *J. Clin. Pharm.*, 3, 43, 1978. With permission.)

(to increase the concentration of undissociated part of an acidic or basic substance), increasing
the salt concentration, or by raising the temperature (increases the saturated vapor pressure).[54]

IV. BIOLOGICAL ASSAYS

By means of biological assays surprisingly low concentrations of certain substances can
be measured, very often with amazing selectivity. In principle the biological system (e.g.,
individuals, cells, enzymes) is exposed to a substance after which some reaction is monitored.

A. LOCAL EFFECTS
Local response in the skin includes changes in the stratum corneum (e.g., barrier function,
elasticity, electrical and spectral properties), in the viable epidermis (e.g., pigmentation,
proliferation), in the dermis (e.g., thickness, blood flow), and in the function of the appendages (e.g., hair growth, sebaceous, and sweat gland excretion). At least in theory, such
biological changes can be measured, although the methods are more qualitative than quantitative. Guy et al. have recently identified and evaluated the various methods employed for
the determination of the bioavailability of topically applied substances.[55] A comprehensive
summary of biological assays used in screening of different topical dosage forms (e.g.,
antimicrobials, antimitotics, antiperspirants, corticosteroids) is found in a review by Haleblian.[56]

1. Vasoconstriction from Steroids
Color changes of the skin, blanching or redness, is undoubtedly the most studied parameter when assessing local effects of penetrants. Topical application of microgram quantities of steroids produce local vasoconstriction, which is visible as blanching. The assay
may be employed both to screen new steroids for clinical efficacy and to determine the
bioavailability of different pharmaceutical formulations.[57-60] The test is carried out in healthy
skin, usually with occlusion of the test area. At a specific time after steroid application a
note can be made whether vasoconstriction is present or not. The blanching can also be
graded and assessed as a function of time. Figure 3 shows the time course of the blanching
for Betnovate cream.[60] Barry has made a comprehensive review of the vasoconstriction assay
and other bioassays for assessments of steroid application.[61] Other agents which also cause
vasoconstriction (e.g., norepinephrine) can be studied in a similar manner.

Attempts have been made to replace the visual assessment by an objective approach. So far, no instrument has been found which is more sensitive, accurate, and reproducible than the visual assessment and at the same time acceptable to volunteers.

2. Erythema

Vasodilatation can be induced by a number of drugs, including nicotinic acid and some of its derivatives.[62] These compounds penetrate the stratum corneum rapidly and cause a distinct erythema on applications to humans. The reddening can be assessed either visually, or by using a noninvasive optical technique, such as laser Doppler velocimetry, LDV.[63-67] The main difference compared to the vasoconstrictor assay is the usually much shorter time of exposure to the drug (minutes instead of hours) and the shorter duration (hours instead of days).

3. Local Anesthesia

Local anesthetics (e.g., lidocaine, mepivacaine, benzocaine) can penetrate the skin and produce anesthesia. One simple method to assess the anesthetic effect is to prick the skin with a needle, and notice the presence or absence of response.[68,69]

B. EFFECTS ON ENZYMES

Toxic organophosphates inhibit the enzyme cholinesterase, whereby the cholinergic nerve transmission is impaired. Methods based on the inhibition of these enzymes have been developed for measuring the penetration of organophosphates both *in vivo* and *in vitro*.[7,70-73]

1. In Vivo

The penetration of organophosphates can be determined on animals by measuring the cholinesterase inhibition in plasma. The organophosphate is applied to the skin, and the decrease in enzyme activity is followed during a period of time (usually hours). The decrease in enzyme activity is then compared with the decrease obtained after injection of known amounts of organophosphate. From the comparison the rate of penetration can be calculated.[71]

2. In Vitro

The *in vitro* method uses a similar approach as the *in vivo* assay. Cholinesterases can be purified from plaice.[74] A standard curve is obtained by partial inhibition of the enzymes by different concentrations of the organophosphate.[7] The degree of inactivation is measured in a spectrophotometer using acetylthiocholine iodide as substrate.[75] The concentration of organophosphate in the receptor fluid can then be determined by adding known amounts of the receptor fluid to the enzyme solution and measuring the degree of inhibition as above.

The *in vitro* enzyme inhibition technique has also been used to study the penetration of a very easily hydrolyzed organophosphate.[73] The experimental set up mimicked the inhibition of cholinesterases in the blood. This could be done by using a flow-through diffusion cell and pumping a cholinesterase solution beneath the skin. From the degree of inhibition the penetration rate of intact compound was calculated. The rate of penetration was also measured by radiochemical detection. The radiochemical detection gave a penetration rate about 50 times higher than the enzymatic assay. The difference between the two results was thought to be due to hydrolysis of the compound in the skin. Thus, by comparing the results derived from different detection methods, additional information about the applied compound (decomposition, purity) may be gained.

ACKNOWLEDGMENT

I would like to thank Vera Franzén, Inger Holmberg, Margareta Johansson, Ulf Jönsson,

Birgitta Olofsson, Agneta Tufvesson-Alm, and Sonja Strandberg for valuable suggestions during preparation of the manuscript.

APPENDIX

A brief list of the major manufacturers of instruments and accessories used in radiochemical analysis is given here.

A. LIQUID SCINTILLATION COUNTERS
Packard Instrument Company 2200 Warrenville Rd., Downers Grove, IL 60515 U.S.

Beckman Instruments, Inc. Nuclear Systems Operation 2500 Harbor Boulevard, Box 3100, Fullerton, California 92634-3100 U.S.

Pharmacia-Wallac Oy, Tietokatu 2, 20101 Turku 10, Finland

B. RADIOISOTOPES
Amersham International plc, Amersham Place, Little Chalfont, Buckinghamshire, England, HP7 9NA

Amersham Corporation, 2636 S. Clearbrook Drive, Arlington Height, Illinois 60005, U.S.

E. I. duPont de Nemours & Co, Biotechnology systems divisions (NEN), 549 Albany Street, Boston, MA 02118, U.S.

ICN Biomedicals, Inc. 3300 Hyland Ave. Costa Mesa, CA 92626, U.S.

MSD Isotopes, Merck Chemical Division, 4545 Oleatha Avenue, St. Louis, MO 62116

C. RADIOSCANNERS
Berthold Radiation Measuring Instruments for Industry, Science and Medicine, Laboratorium Prof Dr. Berthold, D-7547 Wildbad 1, P.O. Box. Germany

Bioscan Inc., 4590 MacArthur Blvd., N.W., Washington, D.C. 20007, U.S.

D. BIOLOGICAL OXIDIZERS
Packard Instrument Company, address see A above.

R. J. Harvey Instrument Corporation, 123 Patterson Street, Hillsdale, New Jersey, U.S.

E. RADIOMETRIC DETECTORS FOR HPLC and GC
HPLC Radioactivity Monitor LB506. Berthold Radiation Measuring Instrument for Industry, Science and Medicine, address see C above.

FLO-ONE/Beta Radioactive Flow Detector, Model IC for HPLC and model GIC for GC. Radiomatic Instruments & Chemical Co., Inc. 5102 S. West Shore Blvd, Tampa, Florida 33611, U.S.

REFERENCES

1. **Evans, E. A.**, *Tritium and its Compounds,* Butterworths, London, 1974.
2. **Geary, W. J.**, *Radiochemical Methods,* John Wiley & Sons, Chichester, New York, 1986.
3. **Kolthoff, I. M. and Elving, P. J.**, Nuclear activation and radioisotopic methods of analysis, in *Treatise on Analytical Chemistry, Part 1 — Theory and Practice,* Vol. 14, Section K, John Wiley & Sons, New York, 1986, chap. 1-8.
4. **Malcolme-Lawes, D. J.**, *Introduction to Radiochemistry,* The Macmillan Press, London, 1979.
5. **Bayly, R. J. and Weigel, H.**, Self-decomposition of compounds labelled with radioactive isotopes, *Nature,* 188, 384, 1960.
6. **Marzulli, F. N., Brown, D. W. C., and Maibach, H. I.**, Techniques for studying skin penetration, *Toxicol. Appl. Pharmacol.,* Suppl 3, 76, 1969.
7. **Lodén, M.**, The in vitro hydrolysis of diisopropyl fluorophosphate during penetration through human full-thickness skin and isolated epidermis, *J. Invest. Dermatol.,* 85, 335, 1985.
8. **Foreman, M. I. and Clanachan, I.**, The percutaneous penetration of nandrolone decanoate, *Br. J. Dermatol.,* 93, 47, 1975.
9. **Lodén, M.**, unpublished data, 1987.
10. **Balke, S. T.**, *Quantitative Column Liquid Chromatography — A Survey of Chemometric Methods,* Journal of Chromatography Library, Vol. 29, Elsevier, Amsterdam, 1984.
11. **Braithwaite, A. and Smith, F. J.**, *Chromatographic Methods,* 4th ed. Chapman and Hall, New York, 1985.
12. **Giddings, J. C., Grushka, E., Cazes, J., and Brown, P. R.**, *Advances in Chromatography,* Vol. 25, Marcel Dekker, New York, 1986.
13. **Heftmann, E.**, *Chromatography: A Laboratory Handbook of Chromatographic and Electrophoretic Methods,* 3rd ed. Van Nostrand Reinhold, New York, 1975.
14. **Katz, E.**, *Quantitative Analysis Using Chromatographic Techniques,* John Wiley & Sons, New York, 1987.
15. **Miller, J. M.**, *Chromatography, Concepts and Contrasts,* John Wiley & Sons, New York, 1988.
16. **Beyermann, K.**, *Organic Trace Analysis,* Ellis Horwood, Chichester, 1984.
17. **Ahuja, S.**, *Ultratrace Analysis of Pharmaceuticals and Other Compounds of Interest,* John Wiley & Sons, New York, 1986.
18. **Frei, R. W. and Zech, K.**, *Selective Sample Handling and Detection in High-Performance Liquid Chromatography,* Journal of Chromatography Library, Vol. 39A, Elsevier, Amsterdam, 1988.
19. **Reid, E.**, *Trace-Organic Sample Handling,* Ellis Horwood, Chichester, 1981.
20. **Schill, G., Ehrsson, H., Vessman, J., and Westerlund, D.**, *Separation Methods for Drugs and Related Organic Compounds,* 2nd ed., Swedish Pharmaceutical Press, Stockholm, 1983.
21. **Poole, C. F. and Schuette, S. A.**, Isolation and concentration techniques for capillary column gas chromatographic analysis, *J. High Res. Chromatogr. Chromatogr. Commun.,* 6, 526, 1983.
22. **Woodget, B. W. and Cooper, D.**, *Samples and Standards,* John Wiley & Sons, New York, 1987.
23. **Krull, I. S. and LaCourse, W. R.**, Post-column photochemical derivatizations for improved detection and identification in HPLC, in *Reaction Detection in Liquid Chromatography,* Krull, I. S., Ed., Marcel Dekker, New York, 1986, chap. 7.
24. **Berridge, J. C.**, *Techniques for the Automated Optimization of HPLC Separations,* John Wiley & Sons, New York, 1985.
25. **Brown, P. A. and Hartwick, R. A.**, *High Performance Liquid Chromatography,* Wiley Interscience, New York, 1988.
26. **Engelhardt, H.**, *Practice of High Performance Liquid Chromatography: Applications, Equipment and Quantitative Analysis,* Springer-Verlag, Berlin, Fed. Rep. Germany, 1985.
27. **Meyer, V. R.**, High-performance liquid chromatographic theory for the practitioner, *J. Chromatogr.,* 334, 197, 1985.
28. **Schoenmakers, P.**, *Optimization of Chromatographic Selectivity. A Guide to Method Development,* Vol. 35, Elsevier, Amsterdam, 1986.
29. **Wainer, I. W.**, *Liquid Chromatography in Pharmaceutical Development,* Aster Publishing, Springfield, 1985.
30. **Snyder, L. R. and Kirkland, J. J.**, *Introduction to Modern Liquid Chromatography,* 2nd ed., John Wiley & Sons, New York, 1979.
31. **Belen'kii, B. G., Gankina, E. S., and Mal'tsev, V. G.**, *Capillary Liquid Chromatography,* Plenum Press, New York, 1987.
32. **Ettre, L. S.**, Selective detection in column chromatography, *J. Chromatogr. Sci.,* 16, 396, 1978.
33. **Scott, R. P. W.**, *Liquid chromatography detectors,* 2nd ed., Vol. 33, Elsevier, Amsterdam, 1986.
34. **Yueng, E. S.**, *Detectors for Liquid Chromatography,* John Wiley & Sons, New York, 1986.

35. **Bashford, C. L. and Harris, D. A.,** *Spectrophotometry and Spectrofluorometry: A Practical Approach,* IRL Press, Oxford, 1987.
36. **Nowicka-Jankowska, T.,** *Comprehensive Analytical Chemistry, Analytical Visible and Ultraviolet Spectrometry,* Vol. 19, Elsevier, Amsterdam, 1986.
37. **Fried, B. and Sherma, J.,** *Thin-Layer Chromatography. Techniques and Applications,* 2nd ed., Chromatographic Science Series, Vol. 35. Marcel Dekker, New York, 1986.
38. **Geiss, F.,** *Fundamentals of Thin-Layer Chromatography,* Huethig Verlag, Heidelberg, Fed. Rep. Germany, 1987.
39. **Roberts, T. R.,** *Radiochromatography — The Chromatography and Electrophoresis of Radiolabelled Compounds,* Elsevier, Amsterdam, 1978.
40. **Zweig, G. and Sherma, J.,** *CRC Handbook of Chromatography,* Vol. 2, CRC Press, Cleveland, OH, 1972, 103.
41. **Lee, M. L., Yang, F. J., and Bartle, K. D.,** *Open Tubular Column Gas Chromatography,* John Wiley & Sons, New York, 1984.
42. **Grob, R. L.,** *Modern Practice of Gas Chromatography,* 2nd ed., Wiley-Interscience, New York, 1985.
43. **Jennings, W. G. and Rapp, A.,** *Sample Preparation for Gas Chromatographic Analysis,* Huethig Verlag, Heidelberg, 1983.
44. **Oehme, M.,** *Hochauflösende Gas-Chromatographie,* Huethig Verlag, Heidelberg, 1987.
45. **Grob, K.,** *On-Column Injection in Capillary Gas Chromatography: Basic Technique, Retention Gaps, Solvent Effects,* Huethig Verlag, Heidelberg, 1987.
46. **Grob, K.,** *Classical Split and Splitless Injection in Capillary Gas Chromatography,* Huethig Verlag, 1986.
47. **Dressler, M.,** *Selective Gas Chromatographic Detectors,* Vol. 36, Elsevier, Amsterdam, 1986.
48. **Chapman, J. R.,** *Practical Organic Mass Spectrometry,* John Wiley & Sons, New York, 1985.
49. **Duckworth, H. E., Barber, R. C., and Venkatasubramanian, V. S.,** *Mass Spectroscopy,* 2nd ed., Cambridge University Press, 1986.
50. **Gaskell, S.,** *Mass Spectrometry in Biomedical Research,* John Wiley & Sons, New York, 1986.
51. **Karesek, F. W. and Clement, R. E.,** *Basic Gas Chromatography-Mass Spectrometry: Principles and Techniques,* Elsevier, Amsterdam, 1986.
52. **Watson, T. J.,** *Introduction to Mass Spectrometry,* Raven Press, New York, 1985.
53. **White, F. A. and Wood, G. M.,** *Mass Spectrometry. Applications in Science and Engineering,* Wiley-Interscience, New York, 1986.
54. **Friant, S. L. and Suffet, I. H.,** Interactive effects of temperature, salt concentration, and pH on head space analysis for isolating volatile trace organics in aqueous environmental samples, *Anal. Chem.,* 51, 2167, 1979.
55. **Guy, R. H., Guy, A. H., Maibach, H. I., and Shah, V. P.,** The bioavailability of dermatological and other topically administered drugs, *Pharm. Res.,* 3, 253, 1986.
56. **Haleblian, J. K.,** Bioassays used in development of topical dosage forms, *J. Pharm. Sci.,* 65, 1417, 1976.
57. **McKenzie, A. W.,** Percutaneous absorption of steroids, *Arch. Dermatol.,* 86, 611, 1962.
58. **McKenzie, A. W. and Stoughton, R. B.,** Method for comparing percutaneous absorption of steroids, *Arch. Dermatol.,* 86, 608, 1962.
59. **Bennett, S. L., Barry, B. W., and Woodford, R.,** Optimization of bioavailability of topical steroids: non-occluded penetration enhancers under thermodynamic control, *J. Pharm. Pharmacol.,* 37, 298, 1985.
60. **Barry, B. W. and Woodford, R.,** Activity and bioavailability of topical steroids. In vivo/in vitro correlation for the vasoconstriction test. *J. Clin. Pharm.,* 3, 43, 1978.
61. **Barry, B. W.,** *Dermatological Formulations. Percutaneous Absorption,* Marcel Dekker, New York, 1983.
62. **Stoughton, R. B., Clendenning, W. E., and Kruse, D.,** Percutaneous absorption of nicotinic acid and derivatives, *J. Invest. Dermatol.,* 35, 337, 1960.
63. **Holloway, G. A.,** Laser Doppler measurement of cutaneous blood flow, in *Non-Invasive Measurements,* Vol. 2, Rolfe, P., Ed., Academic Press, London, 1983, 219.
64. **Guy, R. H., Tur, E., and Maibach, H. I.,** Optical techniques for monitoring cutaneous microcirculation, *Int. J. Dermatol.,* 24, 88, 1985.
65. **Guy, R. H., Wester, R. C., Tur, E., and Maibach, H. I.,** Non-invasive assessments of the percutaneous absorption of methyl nicotinate in humans, *J. Pharm. Sci.,* 72, 1077, 1983.
66. **Ryatt, K. S., Stevenson, J. M., Maibach, H. I., and Guy, R. H.,** Pharmacodynamic measurements of percutaneous penetration enhancement in vivo, *J. Pharm. Sci.,* 75, 374, 1986.
67. **Guy, R. H., Tur, E., Bugatto, B., Gaebel, C., Sheiner, L. B., and Maibach, H. I.,** Pharmacodynamic measurements of methyl nicotinate percutaneous absorption, *Pharm. Res.,* 1, 76, 1984.
68. **McCafferty, D. F., Woolfson, A. D., McClelland, K. H., and Boston, V.,** Comparative in vivo and in vitro assessment of the percutaneous absorption of local anaesthetics, *Br. J. Anaesth.,* 60, 64, 1988.
69. **Broberg, F., Brodin, A., Åkerman, B., and Frank, S. G.,** In vitro and in vivo studies on lidocaine formulated in an o/w cream and in a polyethylene glycol ointment, *Acta Pharm. Suec.,* 19, 229, 1982.

70. **Griesemar, R. D., Blank, H. I., and Gould, E.,** The penetration of an anticholinesterase agent into skin. III. A method for studying the rate of penetration into the skin of the living rabbit, *J. Invest. Dermatol.,* 31, 255, 1958.

71. **Fredriksson, T.,** Studies on the percutaneous absorption of sarin and two allied organophosphorus cholinesterase inhibitors, *Acta Derm. Venereol. Stockh.,* 38 (Suppl. 41), 1958.

72. **van Hooidonk, C., Ceulen, B. I., Kienhuis, H., and Bock, J.,** Rate of skin penetration of organophosphates measured in diffusion cells, in *Mechanisms of Toxicity and Hazard Evaluation,* Holmstedt, B., Lauwerys, R., Mercier, M., Roberfroid, M., Eds., Elsevier, Amsterdam, 1980, 643.

73. **Lodén, M.,** Hudpenetrerande, hudirriterande och hudsensibiliserande egenskaper hos etyl dietylklorfosfat (LG1), FOA rapport, C 40217-C3(C1), National Defense Research Institute, Sweden, 1985.

74. **Lundin, J., Bovallius, Å., Holmberg, L., and Lindner, G.,** Halvskaletekninsk framställning av kolinesteras ur kroppsmuskel från rödspätta (Pleuronectus platessa), FOA 1 rapport, C1330-34 (36), National Defense Research Institute, Sweden, 1969.

75. **Ellman, G. L., Courtney, K. D., Andres, V., and Featherstone, R. M.,** A new and rapid colorimetric determination of acetylcholineesterase activity, *Biochem. Pharmacol.,* 7, 88, 1961.

Chapter 9

DATA INTERPRETATION AND ANALYSIS IN PERCUTANEOUS ABSORPTION STUDIES

William R. Ravis

TABLE OF CONTENTS

I. INTRODUCTION

Most *in vitro* and *in vivo* percutaneous absorption studies include comparisons of the effects of chemical structure, dosing vehicle, or skin conditions on the rate and completeness of topical absorption. Such experiments search to find basic information on means to improve, inhibit, or control the rate of skin penetration and systemic availability of drugs, agents, and toxicants. The development of sound experimental design and conditions becomes critical to the data evaluation and statistical methods irrespective of the degree of sophistication these approaches might employ. Elaborate modeling of *in vitro* or *in vivo* results cannot correct faults in study design or lack of appropriate experiments. This becomes of particular concern in *in vitro* penetration studies where skin selection, thickness, storage, integrity, and preparation techniques can be a source of considerable problems in reproducibility of results.

Numerous mathematical approaches for obtaining estimates for rates of *in vitro* and *in vivo* percutaneous absorption have been introduced.[1-7] Many of the methods are important and useful for understanding topical absorption in the context of skin anatomy and properties. In most cases, such approaches appear mathematically overwhelming to researchers who have been recently introduced to percutaneous absorption studies. Generally, these percutaneous absorption models are valuable for quantitative predictions or simulations of the effects of the physiochemical properties of penetrants and vehicles on relative rates of absorption. However, most experiments do not yield sufficient data and information to properly establish parameters for complex models without making numerous essential assumptions. The following sections will provide an overview of some of the generalized approaches to evaluating percutaneous absorption results with an effort to limit extensive discussion of theoretical modeling and their derivation.

II. DESIGN AND EVALUATION FOR IN VITRO DIFFUSION CELL METHODS

Comparisons of agent penetration and vehicle effects have been extensively studied through the application of diffusion cell techniques.[8] The fact that these types of *in vitro* studies can maintain skin sections with minimal changes in skin permeability characteristics provides a methodology which can be useful in examining physiochemical properties and behavior with significance toward predicting in vivo absorption. Since the stratum corneum, which appears to be the rate limiting barrier, is stable in such experiments, relative *in vitro* percutaneous transfer rates have been viewed as pertinent to *in vivo* absorption. For these reasons, *in vitro* techniques have received wide use and acceptance.

The apparent rate of permeability absorption reflects the diffusion and partitioning of an agent through both the vehicle and the skin tissue. If studies are concerned with comparative differences in skin penetration, the release of the agent from the vehicle should not be rate limiting to the absorption process. Very hydrophobic compounds may be rate limited by the diffusion through donor solutions and the compound's affinity for a lipophilic vehicle or solutions containing solubilizers placed in the donor compartment. Therefore, it may be difficult to evaluate skin penetration for these compounds in the absence of vehicle and solution effects.

In vitro human skin can display a marked variability in agent permeability and this must be carefully considered in the design and evaluation of *in vitro* experiments. It is preferable that all comparative studies be performed with the skin from the same body site and subject. In many study designs, this is not possible and efforts to standardize anatomical site, skin preparation, and storage conditions must be attempted.[9] With cadaver skin sections, the time from subject death, skin harvesting, storage, and use can be critical to percutaneous transfer

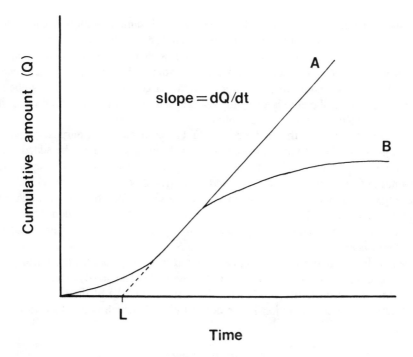

FIGURE 1. Plot of cumulative amount in receptor fluid vs. time for studies with diffusion cells. (A) Infinite dose with steady-state conditions. (B) Depleted or finite dose.

rates thus recording of conditions is required. Performing a repetitive permeability study with the same agent or tritiated water on all specimens may serve to identify specimens and experiments in which skin conditions are inconsistent. Technical problems with tritiated water studies include knowing the specific activity after skin application and the sensitivity of these methods to changes in skin permeability. Ideally, samples of each skin specimen should be represented among all agents and/or conditions being studied. For instance, if six agents are to be compared with respect to their *in vitro* percutaneous absorption, studies with all six should be performed with each new section of skin. This would reduce bias in comparisons of transfer rates and can be factored out as a variable in statistical comparisons.

Diffusion cell conditions can be chosen such that the dose in the donor compartment may be viewed either as infinite or finite in amount. With an infinite dose, the donor compartment would only lose a minimal amount of the dose to tissue equilibrium and transfer to the receiver compartment during the course of the experiment. In this case, the concentration gradient is relatively constant throughout the study. Under finite dose situations the loss of agent from the donor compartment or surface of the membrane results in a significant decrease in the concentration gradient between donor and receiver compartments during the experiment.

A. INFINITE DOSE CONDITIONS

Infinite dose conditions can be assumed if the concentration of the agent in the donor compartment is constant throughout the experiment and the receptor compartment is an effective sink for agent. This latter condition can be maintained with receptor concentrations being less than 5% of the donor concentration during the experiment. Under such conditions, a constant driving force or gradient will be established after an equilibrium period and the rate of appearance of agent in the receptor compartment will be constant. A plot of the cumulative concentration or amount (Q) of agent in the receptor fluid vs. time would follow the description shown in Figure 1. In some cases, this constant gradient can be maintained without the removal of the receptor fluid. Thereby, small samples with time can be removed and replaced with buffer and the total accumulated concentration of agent in the receptor solution at the end of the study is maintained at less than 5% of the donor compartment final concentration. When greater than 5% of the volume is removed and replaced in the receptor, it may be necessary to correct for dilution. Large volume sampling can be avoided by developing analytical methods and using specific activities of tracers which yield adequate detection with small receptor samples volumes.

Under conditions where the total receptor fluid is removed for the purpose of maintaining a constant gradient, the amount of agent in each withdrawn sample can be added to provide a cumulative plot (Figure 1). The removal and replacement of the total receptor fluid has the advantage of providing large volumes and quantities of penetrant for assay. Sample times for exchanges of the receptor solution should be chosen to keep the receptor concentration less than 5% of the donor. If a question exists with respect to the constant donor concentration, the donor cell volume could also be replaced at regular intervals in order to have a near infinite dose system.

A permeability coefficient (P) can be obtained from the slope of the linear portion of the cumulative amounts (Q) vs. time profile[10] (Figure 1). The steady-state slope (dQ/dt) is a function of the permeability coefficient, the surface area (S) of the skin exposed to the donor fluid, and the concentration in the donor fluid (C) which is assumed to be constant (Equation 1).

$$dQ/dt = PSC \tag{1}$$

Therefore, the permeability constant equals (dQ/dt)/(S C). Equation 1 follows Fick's First Law of Diffusion.

The diffusion coefficient for agent within the skin is reflected in the lag time (L) which represents the time required for the constant gradient to develop from the donor compartment, across all skin barriers, and to the receptor fluid. Due to the complexities of diffusion through skin barriers, the diffusion coefficient (D) obtained in this matter is an apparent diffusion parameter. Still, this parameter is an useful comparative value when all conditions of the experiments are maintained uniform. If the extrapolated time intercept (L) can be estimated, a diffusion coefficient is obtained as:

$$D = h^2/(L6) \tag{2}$$

where h is membrane or skin thickness.

Based on the above method for estimating diffusion and permeability coefficients, several difficulties become apparent. The decision as to which data should to be utilized in the calculation of the steady-state slope and the lag time intercept can be subjected to bias. Plots which do not approach linearity can result from prolonged lag time periods or barrier decay during the experiments (Figure 1). Considerably more variability will be observed in parameters when a suitably linear portion cannot be evaluated over a reasonable time. An attempt to achieve more consistent estimates of diffusion and permeability coefficients which utilizes all Q values instead of just the steady-state values has been proposed by Foreman and Kelly.[11]

$$Q = \frac{MSCDt}{h} - \frac{ShMC}{6} - \frac{2ShMC}{\pi^2} \sum_{n=1}^{i} \frac{(-1)^n}{n^2} \exp(-Dn^2\pi^2 t/h^2) \tag{3}$$

In Equation 3, C is the assumed constant donor concentration and M is the solvent-membrane (solvent-skin) partitioning coefficient. Q values can be evaluated as a function of time by nonlinear least-squares methods to obtain M and D with h and C as known values. The last term has been dropped when it adds less than 0.1% to the Q value. To relate this approach to the estimation of P and D from constant slope methods, Equation 3 can be simplified during the steady-state linear portion where dQ/dt or the slope is constant. At these time periods, only the first two terms contribute to the value of Q. Thereby, P of Equation 1 appears equal to D M/h. Thus, M can be estimated from D which is obtained from the lag time intercept of the linear plot and a known value of h. How well estimates of M, D, and P agree between steady-state methods (Equations 1 and 2) and methods in which all data is evaluated (Equation 3) depends on how well the experimental results follow the classic situation represented in Figure 1.

B. FINITE DOSE CONDITIONS

With some cell and experimental systems, the donor concentration may not be viewed as constant and a steady-state condition would not occur. This might occur when an agent in the donor buffer is very hydrophobic and the agent rapidly partitions into and through the skin membrane decreasing the donor compartment concentration. When a dose is applied in a vehicle on the skin surface and the vehicle evaporates or remains on the surface, a significant amount of the agent may diffuse into the skin and receptor with time and a finite dose situation may arise. In these cases, the donor concentration and therefore, the concentration gradient decreases with time. It is still important to maintain the receptor compartment as an infinite sink (receptor concentration <5% of donor) by either removing the volume of this compartment or by employing a constant flow through the receptor portion of the cell. Evidence of a situation in which the donor concentration is significantly depleted is shown in the Q vs. time plot (Figure 1, Curve B). As the experiment continues, dQ/dt or the slope decreases with time due to the decrease in the initial donor concentration.[12]

After time for tissue uptake and for gradients to be established in the skin membranes and tissues, the amount of agent in the donor area (Qd) would decline according to a first-order process as described by Equation 4.

$$\frac{dQd}{dt} = KSQd = PSQd/Vd \qquad (4)$$

Integration of Equation 4 provides Equation 5.

$$\ln(Qd/Qd0) = KSt = PSt/Vd \qquad (5)$$

A plot of the natural logarithm of the ratio of Qd at each sampling time divided by the initial amount (Qd0) vs. time would appear linear as defined by Equation 5 and shown in Figure 2.[7] The slope is equal to K S where K is equal to P divided by the volume of the donor compartment (Vd). P can be estimated when the donor compartment contains a fluid volume. However, in the case where the dose is applied as a film on the skin, values of K can be compared and an estimate of P cannot be obtained.

Equation 5 applies when the amount in the donor compartment (concentration times Vd) is monitored with time. An estimate of K or P can also be made by observing the rate of appearance of agent in the receptor (dQr/dt). After a gradient is generated across the skin membrane, dQr/dt may be proportional to dQd/dt. Thus, the linear portion of a plot according to Equation 6 will yield a slope equal to K S or P S.

$$\ln(dQr/dt) = PSt + intercept = KSt + intercept \qquad (6)$$

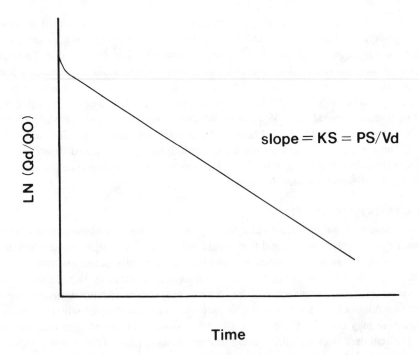

FIGURE 2.   Plot of natural logarithms for the ratio of the amount in donor fluid/initial amount for finite dose studies.

In the equation, t should be the midpoint of the time interval. Estimates of P and K would be expected to be similar for dQd/dt and dQr/dt data, as long as the rate-limiting step for the appearance of the compound in the receptor compartment is the passage of compound from the donor fluid into the skin and not slow release of the compound from the skin tissue to the receptor fluid. If the agent is slowly released from the tissue, it will accumulate in the skin tissue and the value of P or K from a plot of ln(dQd/dt) will be greater than that of ln(dQr/dt) or ln(Qd/Qd0).

Due to the complexity of skin tissue and barriers, it may be difficult to relate parameters and results from finite to infinite dose studies. Since infinite dose experiments usually represent overall skin tissue penetration, permeability parameters obtained by monitoring appearance of the compound in the receptor solution for finite dose conditions might be expected to relate better than disappearance data (dQd/dt) to those obtained from infinite dose studies. The disappearance of agent from the donor compartment reflects not only skin penetration but also tissue equilibrium.

C. ASSESSMENT OF *IN VITRO* METABOLISM

Several chemicals and drugs have been shown to be subject to metabolism during their passage through the epidermal tissues. Skin metabolism can be studied *in vitro* by microsomal preparations, homogenates, and diffusion cell systems. With intact skin and topical exposure, a compound would pass through the stratum corneum followed by possible metabolism of a portion of the compound to metabolites as it diffuses through the epidermal and dermal layers. Formed metabolites may then diffuse and partition into the dermal vasculature or back to the stratum corneum and donor fluid. Partitioning and diffusion of metabolites may be different than the parent compound. Therefore, the appearance rate of metabolites in the receiver compartment is a complex function with the ratio of metabolites to parent compound changing with time.

Skin metabolism, as examined with diffusion cells, is probably best observed under infinite dose or constant exposure conditions. As long as the tissue remains viable and metabolites do not accumulate to any significant extent in the donor compartment, a steady-state flux of metabolite and parent compounds may be noted. During this steady-state period, a constant ratio of metabolites to parent may be found and the appearance rate of metabolite in the receiver fluid may be rate limited by either the metabolic rate or the diffusion rate of compound to the epidermal tissues. If back diffusion to the donor compartment is insignificant, the ratio of metabolite to parent compound during steady-state conditions may be useful in comparing metabolic rates among compounds. To evaluate the saturability of skin metabolic pathways, the concentration of agent in the donor fluid might be varied over a severalfold range.

III. DESIGN AND EVALUATION FOR IN VIVO METHODS

The rate and completeness of percutaneous absorption can be evaluated in vivo through the collection of plasma and urinary data, monitoring amounts of penetrant remaining on the skin, and by observing physiological response. Physiological responses such as localized blanching following corticosteroids[13,14] or nicotinates[15] can be applied to comparisons of rates and extents of skin and systemic penetration for varying vehicles and topical formulations. While assessment of physiological, pharmacological, and toxicological effects appear to be the most direct, most agents do not express such effects in a quantifiable enough manner to permit useful comparisons of percutaneous absorption differences. Monitoring the surface disappearance of a radiotracer by a counter placed on the application area seems practical and straightforward. However, results of such studies may be ambiguous due to quenching and variations in quenching between application sites and subjects. As a result, the methodologies most widely applicable to the study of *in vivo* percutaneous absorption are evaluations based on the time course of the agent in plasma or collected urine. Since systemic and urinary concentrations following topical application may be very low, studies of these types may be limited to agents that are available as radiotracers or for which sensitive assay methods are available.

Following topical exposure, plasma concentrations or urinary levels of an agent are not only an expression of the degree of percutaneous absorption but also a function of the distribution and elimination characteristics of the agent in the body. For this reason, a simple comparison of a plasma or urine concentration at some time after exposure is not adequate to evaluate absorption differences between different compounds. Different compounds will have unique elimination and distribution properties. In the case of contrasting the *in vivo* topical absorption of a series of compounds, it may be necessary to separately characterize the disposition parameters of an agent by separate studies in order to extract the absorption information from the time course of concentrations in biological samples. Problems in interpretation of results even with disposition information can occur when metabolic pattern changes with route of administration.[16] The elimination and distribution features of a compound can be disregarded when studies are performed with the same compound applied topically as in the case of examining the effects of vehicle, application area, etc., on the percutaneous absorption of one particular agent. However, even in such cases, it must be recognized that different subjects will display variation in their disposition parameters.

A. CONSTANT DOSE EXPOSURE
1. Plasma Concentration Studies
In vivo constant exposure or infinite reservoir experiments are analogous to the infinite *in vitro* dose situation in that there is a constant gradient of agent across the skin. A constant gradient might be maintained by a delivery device or by a topical vehicle which maintains a constant concentration of agent available for absorption. To study skin absorption and not

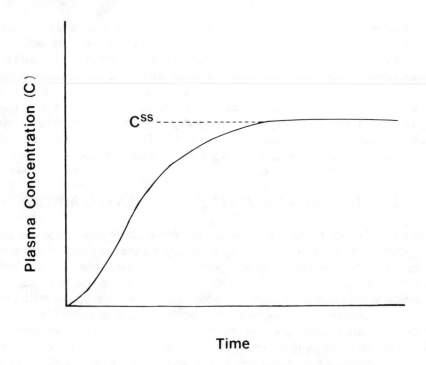

FIGURE 3. Plasma concentration vs. time for an infinite *in vivo* dose.

vehicle or device release, the rate-limiting step to absorption should not be the release from the donor component or media. As for *in vitro* experiments, a time lag may be apparent. For *in vivo* experiments, time is required for a gradient to be established from the source, across the viable tissue, and to the blood of the capillaries which removes agent from the skin area. An expected time course for the concentration of the agent in the plasma is shown in Figure 3. The time required to achieve a steady-state concentration in the plasma is not only dependent on the topical lag time but also the $t_{1/2}$ of the agent in the body. This total time would be the absorption lag time plus approximately five disposition half-lives for the specific compound. Excluding absorption lag time, 5 disposition half-lives alone are required to achieve 97% of the Css value. The steady-state plasma concentration of the compound (Css) will be a function of the zero-order absorption rate (A) and the total body clearance of the compound (Cl) (Equation 7).

$$Css \ = \ A/Cl \ = \ PSC/Cl \eqno(7)$$

The Cl represents the removal rate of agent from the plasma by all eliminating organs and tissues and is expressed in units of volume of plasma per unit time.

The apparent absorption rate under ideal circumstances would be expected to be equal to the product of the permeability constant, the surface area, and the agent concentration at the topical site. Therefore, A could be correlated with *in vitro* constants. Lag times would be more difficult to compare. While theoretically the *in vivo* absorption lag time could be estimated knowing the $t_{1/2}$ of the agent, it would be difficult to assess accurately the time that steady-state is achieved without computer simulations and fitting.

In studies where the absorption rates of the same agent at the same dose but with different vehicles and experimental skin conditions are to be contrasted, Css values can be compared and would be proportional to A. However, for comparisons of the percutaneous absorption of different compounds as perhaps from a series of compounds, Cl varies for each compound

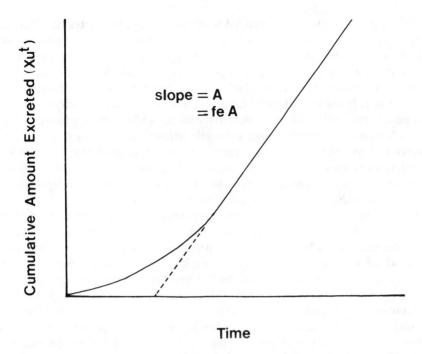

FIGURE 4. Cumulative amount excreted (Xu^t) vs. time for an infinite *in vivo* dose.

and the Css cannot be viewed as proportional to the compound's absorption rate. In such cases, Cl may need to be estimated for each compound by separate intravenous pharmacokinetic studies with the assumption that metabolite patterns do not change with route of administration. In the presence of saturable percutaneous metabolism, the completeness of skin absorption may be dose-dependent and reflected in disproportionate increases in Css and A with increasing dose. Measurements of total radioactivity in the plasma may be suitable when the same agent is being examined but not appropriate when different agents are being compared. In the case of total radioactivity measurements, the time required to reach a steady-state concentration of radioactivity in the plasma reflects the absorption lag time and either the parent or metabolite $t_{1/2}$'s depending on which one is greater.

2. Urinary Studies

The collection of urinary data under constant dose conditions has some advantages for examining *in vivo* percutaneous absorption compared to plasma results. For many compounds, the rate and completeness of percutaneous absorption in combination with the disposition characteristics of the compound may yield low levels of agent and metabolites in the plasma. This may limit the usefulness of plasma concentration methods. In addition, plasma collection requires cannulation and blood sampling whereas urinary studies require only the collection of urine at designated times after exposure.

A typical cumulative amount excreted (Xu) vs. time plot (Figure 4) displays a lag time followed by a linear region which is proportional to the absorption rate (A).[4] The slope is equal to the fraction of the dose that appears in an analytically detectable form in the urine times the absorption rate. In the case of measuring only parent compound, the slope is the fraction of the agent excreted unchanged in the urine (fe) times A and Xu^t vs. time can be described by Equation 8.

$$Xu^t = feA(t - tl - t1/2\ 1.443) \qquad (8)$$

In Equation 8, t1 is the percutaneous absorption lag time for parent drug.

When total radioactivity is monitored in the urine, A can be obtained following some assumptions. For studies of total radioactivity, the fraction might be assumed to be equal to 1 with the slope directly representing the absorption rate. Using excretion rates of total radioactivity in the urine as estimates of A provides a simple way to estimate absorption rates and has advantages over the plasma concentration methods. As long as the entire absorbed dose is found as total radioactivity in the urine and does not vary from compound to compound, cumulative urinary radioactivity methods allow the comparison of different compounds without concern for pharmacokinetic differences. Should a series of compounds be known to have biliary or fecal elimination as important pathways, the fraction of the total radioactivity ever excreted in the urine will be less than 1 and may differ among compounds of the series. This could limit the direct comparisons of excretion rates of total radioactivity in evaluating absorption rates among compounds.

The lag time (Figure 4) is a function of the percutaneous absorption lag time (t1) and the half-lives of the agent or agent and its metabolites. This overall lag time has been referred to as a pharmacokinetic lag time.[4] If only parent agent is detected (Equation 8) and the pharmacokinetics of the agent can be described by a simple one-compartment model, the pharmacokinetic lag time would equal the lag time for percutaneous absorption plus the agent's $t_{1/2}$ times 1.443 (t1 + $t_{1/2}$ 1.443). The value of $t_{1/2}$ 1.443 is equal to the reciprocal of the compound's elimination rate constant. If the $t_{1/2}$ is known from intravenous studies, the absorption lag time can be estimated. The lag time of such plots may not be useful to compare t1 values for different agents unless the $t_{1/2}$ is known and can be separately subtracted. When total urinary radioactivity is determined for a group of compounds, the total radioactivity $t_{1/2}$ will be a complex function which may still be estimated from intravenous studies and subtracted from lag time as long as the metabolic pattern for the agent is not dependent on route of administration.

B. FINITE DOSE: PLASMA AND URINARY DATA

Whereas the constant exposure or infinite dose situation yields a steady state plasma concentration and urine excretion rate of agent due to a constant absorption gradient, plasma concentration and excretion rates after a finite topical dose continuously vary with time as the absorption gradient and absorption rate (A) diminishes. For this reason, the evaluation of finite dose results becomes more complex than that of the infinite dose case. Since renal excretion rates are equal to the agent's renal clearance times the plasma concentration, the time course of the absorption and disposition processes can be followed from the excretion rates as with plasma concentrations. As shown in Figure 5, the concentrations and excretion rates will increase to a maximum value then decline. The terminal logarithmic decline may represent the agent's disposition $t_{1/2}$ when the compound is rapidly absorbed or the absorption kinetics in cases where absorption is slow relative to elimination.[17] To identify which situation exists requires intravenous dosing and comparison of terminal $t_{1/2}$s.

In recent years, several models have been proposed to predict and evaluate the influence of the percutaneous absorption process on the time course of drugs and toxicants and their metabolites in plasma and urine.[1,3-6] Some models have considered the gradients of the agent that occur across skin layers as well as the effects of vehicle-skin partitioning and thickness of the applied vehicle on *in vivo* absorption rates. These types of models utilize the nonsteady-state equations of Fick's second law.[3] Other approaches have described agent penetration, partitioning, and skin accumulation in terms of first-order constants.[1,5,6,18] Modeling with first-order constants has been applied towards predicting plasma or urinary levels of drugs following finite doses.[18] However, this approach requires knowledge of *in vitro* constants such as diffusion coefficients in order to simulate *in vivo* data and is not useful to directly evaluate plasma and urinary data. While these models are very important toward understanding the influence of physiochemical conditions and aspects on percutaneous absorption,

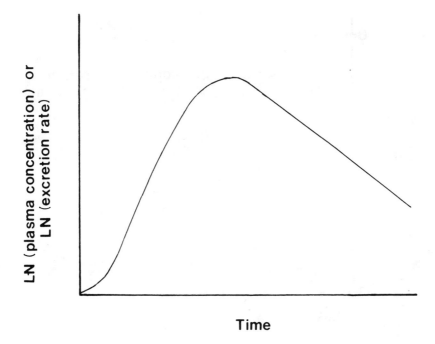

FIGURE 5. Plot for the natural logarithms of plasma concentration (C) or renal excretion rate (ER) following a finite *in vivo* dose.

their application to data reduction and interpretation may be beyond the needs of most studies of topical absorption. For this reason, the methods presented are more generalized and are aimed toward direct estimating and comparing rate and completeness of *in vivo* absorption.

Both the rate and extent of in vivo absorption can be assessed from measurements of the area under plasma concentration curve (AUC) or the amount of compound excreted in the urine (Xu). Values of AUC and Xu represent cumulative measurements and, for this reason, are superior to single time point comparisons of plasma and urinary data. By comparing AUC and Xu values for the same agent among different routes, topical vehicles, or skin conditions, estimates of the completeness or relative completeness of the topical absorption can be made. AUC estimates, in most cases, are made by use of the trapezoidal rule and requires that plasma samples be taken at times which will appropriately characterize the plasma concentration time course. An estimate of the overall or absolute extent (F) of percutaneous absorption can be made by comparing the AUC^{∞} or Xu^{∞} from 0 to infinity from intravenous (i.v.) dosing to that obtained following topical administration (Equation 9).

$$F = \frac{(AUC^{\infty}_{topical})(Dose_{iv})}{(Dose_{topical})(AUC^{\infty}_{iv})} = \frac{(Xu^{\infty}_{topical})(Dose_{iv})}{(Dose_{topical})(Xu^{\infty}_{iv})} \tag{9}$$

Ideally, both routes would be performed in the same subject but averages from separate groups of subject may be more practical. The completeness of absorption can also be examined from the total cumulative amount of agent and metabolites which would appear in the urine. Comparisons of total radioactivity in the urine for estimating F after i.v. and topical doses are suitable when skin metabolism is insignificant or interest is in total percutaneous availability of an agent and percutaneously formed metabolites. In the presence of suspected skin metabolism,[16,19] the parent compound's AUC^{∞} or Xu^{∞} for intravenous and topical routes can be contrasted to assess the "first-pass" effect of the skin on systemic

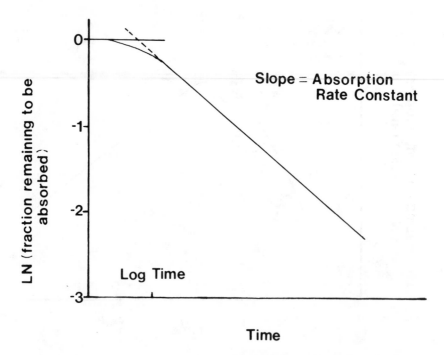

FIGURE 6.   Plot of the natural logarithms for fraction remaining to be absorbed for plasma or urinary data. Linear section represents first-order absorption with a slope equal to the first-order absorption rate constant.

absorption. A method for estimating first-pass skin metabolism without i.v. administration have been proposed.[20] By this method, agent was applied to the volar surface of the wrist and plasma concentrations or AUCs of agent and metabolites for samples drawn from the contralateral and the ipsilateral antecubial forearm were compared.

If comparative effects of vehicles or penetration enhancers on the percutaneous absorption of a single agent are desired, the AUCs for the same agent and dose can be compared to a control or test situation without need for intravenous data. On the other hand, when different compounds are being compared, the pharmacokinetics of each compound would need to be established and F calculated. Differences in Cl and fe among compounds will influence AUC's and total urinary excretions. As in the case of infinite doses for comparing groups of compounds, complete collection of urine and determination of total radioactivity can be used to estimate overall extent of absorption. However, this approach assumes all absorbed compound will appear in the urine as parent or metabolites with none eliminated by other routes.

While the total AUC (AUC^{∞}) represents total extent of absorption, the AUC up to a particular time (AUC^{t}) contains information with regard to the rate of systemic absorption. For compounds whose disposition kinetics can be adequately described by a simple one-compartment model, the amount of compound absorbed can be expressed as the amount that has been eliminated plus the amount in the body. Based on this approach which has been described by Wagner and Nelson[21] for a compound with one-compartment features, the fraction which remains to be absorbed can be predicted by Equation 10.

$$\text{fraction remaining to be absorbed} = 1 - \frac{C^{t} + (0.693/t1/2)(AUC^{t})}{(0.693/t1/2)(AUC^{\infty})} \tag{10}$$

The AUC^{t} is the AUC from time $= 0$ to t and C^{t} is the plasma concentration of the agent at time t. A plot of the natural logarithm of fraction remaining to be absorbed vs. time will display a lag time and then decline according to the order of the absorption process. Figure 6 shows a profile of the fraction remaining to be absorbed for a compound which displays first-order percutaneous absorption. For a first-order absorption process, the linear slope of the natural logarithm plot is equal to the absorption rate constant, ka, in units of 1/time. *In vivo* lag times could be estimated by extrapolating the terminal slope to a time when the fraction remaining to be absorbed is 1 (LN 1 = 0). Similar plots can be generated with urinary data as shown in Equation 11.

$$\frac{\text{fraction remaining}}{\text{to be absorbed}} = 1 - \frac{(ER^t) + (0.693/t1/2)(Xu^t)}{(0.693/t1/2)(Xu^\infty)} \tag{11}$$

For urinary data, ER equals the excretion rate (dXu/dt) in amount per unit time. The use of these methods requires an estimate of the disposition $t_{1/2}$ and the total AUC^∞ or Xu^∞. When $t_{1/2}$ and AUC^∞ or Xu^∞ cannot be obtained, estimates of Cl or fe will be needed to predict the fraction of the agent which has been absorbed.

In the context of *in vitro* skin penetration parameters for compounds, percutaneous absorption from a finite dose would be expected to display first-order characteristics with the absorption rate constant, ka, equal to the K S or P S/Vd of *in vitro* studies. This would be observed in the absence of vehicle or device release effects. Due to the complex events involved in *in vivo* absorption, rank order correlations between *in vitro* and *in vivo* results might be anticipated while exact predictions of *in vivo* rates by *in vitro* rates is not likely. Should the order of the absorption process be complex or not distinguishable, differences in rates of absorption for the same penetrant with varying conditions could be compared from observed AUC^t and Xu^t at specific times after exposure. However, when values are contrasted at only one time, the rank order of the results could depend on the time period chosen. Comparisons of times of peak plasma concentration or peak excretion rate can provide information on the relative order for rates of percutaneous absorption for the same compound. The time of peak reflects a lag time and rate of absorption whereas the peak value itself is a function of the extent of absorption as well as the rate of absorption.

An approach for contrasting rates of absorption which has the advantages of being independent of absorption order and disposition kinetic models is evaluation by mean residence times (MRT).[22] The mean residence time for disposition (MRT_{iv}) is the average time an agent molecule remains in the body following an intravenous bolus dose. Following topical administration with systemic absorption, the overall mean residence time ($MRT_{topical}$) for an agent will be the MRT_{iv} plus the mean residence time of percutaneous absorption (MRT_{abs}). The MRT_{abs} is the average time required for an agent molecule to disappear from the application site. Therefore, the MRT_{abs} reflects both the lag time and the absorption rate. MRT is calculated from the AUC^∞ and the first-moment of the AUC curve ($AUMC^\infty$). $AUMC^\infty$ can be obtained by utilizing the trapezoidal rule with the values of C times t as a function of t. $MRT_{topical}$ and MRT_{iv} equals $AUMC^\infty/AUC^\infty$. When different agents are being compared and MRT_{iv} is determined from separate studies, MRT_{abs} can be estimated as $MRT_{topical} - MRT_{iv}$. If studies are performed with the same compound and subjects are the same for all studies or fairly uniform in their disposition kinetics, comparisons of $MRT_{topical}$s can be made directly. The disadvantage of the MRT approach is that areas must be estimated to infinity requiring fairly complete sampling with extrapolation of terminal areas from the last measured point. MRT estimates can also be obtained from urinary data except that the moment of the ER curve vs. time is constructed from ER times the midpoint time of the interval. The MRT is calculated then as the total ER times t integral divided by the Xu^∞.

III. STATISTICAL EVALUATION

In percutaneous absorption studies, there is interest in determining whether differences exist in rate and completeness of absorption among compounds or experimental conditions.

To compare a specific condition or compound to others with confidence, all other experimental conditions of the study must be the same. This would mean the same skin region, subject, vehicle, temperature, device, etc. Under such design, parameters derived from percutaneous absorption experiments can be compared and examined for significant differences by applying a *t*-test.[23] For testing two means when the variance in the mean values is estimated from the sample, the *t*-test is

$$t = \frac{\text{Mean}_1 - \text{Mean}_2}{\sqrt{\dfrac{(n_1 - 1)SD_1 + (n_2 - 1)SD_2}{n_1 + n_2 - 2}} \sqrt{\dfrac{1}{n_1} + \dfrac{1}{n_2}}} \tag{12}$$

If t is greater than the t value for $n_1 + n_2 - 2$ degrees of freedom from the Student's *t* table, significance is accepted at the decided confidence limit. SD and n are the standard deviation and number of observation for each group, respectively. Application of the *t*-test assumes that the parameter is normally distributed, sample groups are independent, and the variance is the same for each group compared. When variance differences between sample groups are great (ratio of variances greater than 3) then nonparametric procedures mentioned below may be recommended.

The *t*-test may be viewed as a simplified form of an analysis of variance (ANOVA). ANOVA, in its many forms, permits the comparison of several means and values representing varying experimental conditions. ANOVA procedures can be performed on a desktop computer with such programs as PC-SAS.[24] When comparing more than two means, ANOVA is preferred over multiple *t*-tests since it utilizes a common variance established from all studies to detect the significant differences among means. An ANOVA approach is considered a powerful tool; however, design is critical to its appropriate application. By expanding to a multiple factor type ANOVA, the effects of separate variables or conditions on an estimated parameter can be evaluated and significance judged. For example, in *in vitro* studies with cadaver specimens, it may not be possible for all proposed studies to be performed with tissue from the same subject and the same area skin. By distributing a portion of each specimen among all experimental conditions that will be compared, it becomes possible to consider the contributions of specimen differences on the measured means. In a case where a series of compounds are being studied, a two-way ANOVA could examine for significant differences in the permeability coefficients among compounds as well as the influence of the specimen source on the coefficients. For studies in which the absorption of several agents is evaluated at more than one donor concentration, a three-factor ANOVA including specimen, compound, and concentration effects could be performed. Following ANOVA procedures, comparisons of treatment or effect means can be made by multiple comparison testing methods.[25]

An argument can be made for evaluating data which represents the desired information in its most basic form with limited data manipulation. An example would be to simply compare amounts absorbed up to some time from *in vitro* or *in vivo* systems instead of comparing rate constants which requires the assumption of a model or order for the absorption process. Values for estimated parameters which reflect rates of agent transfer can be very dependent on the model assumed and utilized in the data evaluation. It is possible that different conclusions may be reached depending on the model and methods chosen. Another statistical consideration which arises with rates constants derived from models concerns the normality of their distribution around a mean value. Since several of the above mentioned assumptions are common to the popular statistical tests, basic comparisons of raw data might be preferable to model dependent constants for these tests.

ANOVA and the *t*-test assume homogeneity in variance and normality. In the process of calculating parameters from raw data, parameters may be obtained of which their normality cannot be fully judged based on a small number of observations. When heterogeneity in variance is identified by such procedures as a F-test of variances,[26] data transformation may be helpful in providing values which display better normality and uniform variance. The most common transformation might involve evaluating logarithms of parameters while other approaches utilize square root, arc sine, or reciprocal transformations. For instance, instead of comparing percentages absorbed directly, the logarithms of the percentages absorbed might be compared. To avoid debates concerning the statistical structure of parameters and data, nonparametric or rank-order methods such as the Wilcoxon rank test[27] or the Kruskal-Wallis test may be preferred. These methods do not assume that values are normally distributed around a mean and they do not require similar variances for samples. Since the statistical features and characteristics of absorption rates and parameters are not fully understood and simple statistical methods may not be applicable across all complex percutaneous absorption models, it is difficult to suggest any particular approach to data reduction and evaluation. Further descriptions and discussions of parametric and nonparametric procedures are found in several statistical references.[23-28]

Current advances in *in vitro* and *in vivo* methodologies and modeling of percutaneous absorption have been extensive. This chapter serves simply as an overview of some generalized and practical aspects for data evaluation of percutaneous absorption results. Through better understanding of percutaneous absorption, in such areas as pathways, metabolism, and reservoir effects, improved approaches to studying specific factors which are important to topical absorption will become apparent.

SYMBOLS

A	=	zero-order absorption rate = P S C
AUC^t	=	area under the plasma concentration curve up to time t
AUC^∞	=	area under the plasma concentration curve up to infinity
$AUMC^\infty$	=	area under curve for the product of c times t vs. time
C	=	concentration in donor compartment
C^t	=	plasma concentration at time t
Css	=	steady-state plasma concentration
Cl	=	total body clearance of compound
D	=	diffusion coefficient
$Dose_{topical}$	=	dose applied topically
$Dose_{iv}$	=	dose administered intravenously
ER^t	=	excretion rate of unchanged drug in urine at time t = dXu/dt
F	=	the extent of absorption
fe	=	fraction of unchanged drug excreted in the urine
h	=	membrane or skin thickness
K	=	P/Vd
L	=	*in vitro* lag time
M	=	solvent-membrane(solvent-skin) partitioning coefficient
MRT	=	mean residence time—average time a molecule is in the body
MRT_{iv}	=	average time a compound is in the body after an intravenous bolus
MRT_{abs}	=	the average time for a compound molecule to be absorbed
$MRT_{topical}$	=	average time for a compound molecule to be absorbed and eliminated from the body = $MRT_{iv} + MRT_{abs}$
P	=	permeability coefficient
Q	=	cumulative amount in receptor compartment
dQ/dt	=	steady-state rate of change in amount in receptor compartment
Qd	=	amount in donor compartment

Qd0	=	initial amount in donor compartment
dQd/dt	=	rate of change in the amount in the donor compartment
Qr	=	amount in receptor compartment
dQr/dt	=	rate of change in the amount in the receptor compartment
S	=	surface area
t	=	time
$t_{1/2}$	=	disposition half-life
t1	=	in vivo percutaneous absorption lag time
Vd	=	volume of donor compartment
Xu^t	=	cumulative amount excreted in urine at time t
Xu^∞	=	cumulative amount excreted in the urine at infinity

REFERENCES

1. **Guy, R. H. and Hadgraft, J.,** Prediction of drug disposition kinetics in skin and plasma following topical administration, *J. Pharm. Sci.,* 73, 883, 1984.
2. **Zatz, J. L.,** Percutaneous absorption: Computer simulation using multicompartmented membrane models, in *Percutaneous Absorption: Mechanisms-Methodology-Drug Delivery,* Bronaugh, R. L. and Maibach, H. I., Eds., Marcel Dekker, New York, 1985, chap. 13.
3. **Cooper, E. R. and Berner, B.,** Finite dose pharmacokinetics of skin penetration, *J. Pharm. Sci.,* 74, 1100, 1985.
4. **Berner, B.,** Pharmacokinetics of skin penetration, in *Transdermal Delivery of Drugs,* Vol. 2, Kydonieus, A. F. and Berner, B., Eds., CRC Press, Boca Raton, 1987, chap. 9.
5. **Guy, R. H. and Hadgraft, J.,** Pharmacokinetics of percutaneous absorption with concurrent metabolism, *Int. J. Pharm.,* 20, 43, 1984.
6. **Guy, R. H. and Hadgraft, J.,** Mathematical models of percutaneous absorption, in *Percutaneous Absorption: Mechanisms-Methodology-Drug Delivery,* Bronaugh, R. L. and Maibach, H. I., Eds., Marcel Dekker, New York, 1985, chap. 1.
7. **Flynn, G. L., Smith, W. M., and Hagen, T. A.,** In vitro transport, in *Transdermal Delivery of Drug,* Vol. 1, Kydonieus, A. F. and Berner, B., Eds., CRC Press, Boca Raton, 1987, chap. 4.
8. **Franz, T. J.,** Percutaneous absorption. On the relevance of in vitro data, *J. Invest. Dermatol.,* 64, 190, 1975.
9. **Pitman, I. H. and Rostas, S. J.,** A comparison of frozen and reconstituted cattle and human skin as barriers to drug penetration, *J. Pharm. Sci.,* 71, 427, 1982.
10. **Scheuplein, R. J.,** Mechanism of percutaneous absorption. I. routes of penetration and the influence of solubility, *J. Invest. Dermatol.,* 45, 334, 1965.
11. **Foreman, M. I. and Kelly, I.,** The diffusion of nandrolone through hydrated human cadaver skin, *Br. J. Dermatol.,* 95, 265, 1976.
12. **Hadgraft, J.,** Variables associated with a kinetic analysis of skin penetration, in *Pharmacology and the Skin,* Vol. 1, Shroot, B. and Schaefer, H., Eds., Karger, Basel, 1986, pp. 154.
13. **McKenzie, A. W. and Stoughton, R. B.,** Method for comparing percutaneous absorption of steroids, *Arch. Dermatol.,* 86, 608, 1962.
14. **Barry, B. W. and Woodford, R.,** Comparative bio-availability and activity of proprietary topical corticosteroid preparations: vasoconstrictor assays on thirty-one ointments, *Br. J. Dermatol.,* 93, 563, 1975.
15. **Stoughton, R. B., Clendenning, W. E., and Kruse, D.,** Percutaneous absorption of nictonic acid derivatives, *J. Invest. Dermatol.,* 35, 337, 1960.
16. **Wester, R. C. and Maibach, H. I.,** Cutaneous pharmacokinetics: 10 steps to percutaneous absorption, *Drug Metab. Rev.,* 14, 169, 1983.
17. **Riegelman, S.,** Pharmacokinetics: pharmacokinetic factors affecting epidermal penetration and percutaneous absorption, *J. Clin. Pharmacol. Ther.,* 16, 873, 1974.
18. **Guy, R. H., Hadgraft, J., and Maibach, H. I.,** Percutaneous absorption in man: a kinetic approach, *Toxicol. Appl. Pharmacol.,* 78, 123, 1985.

19. **Wester, R. C. and Maibach, H. I.**, In vivo animal models for percutaneous absorption, in *Percutaneous Absorption: Mechanisms-Methodology-Drug Delivery,* Bronaugh, R. L. and Maibach, H. I., Eds., Marcel Dekker, New York, 1985.
20. **Yacobi, A., Baughman, R. A., Cosulich, D. B., and Nicolau, G.**, Method for determination of first-pass metabolism in human skin, *J. Pharm. Sci.,* 73, 1499, 1984.
21. **Wagner, J. G. and Nelson, E.**, Kinetic analysis of blood levels and urinary excretion in the absorptive phase after single doses of drug, *J. Pharm. Sci.,* 53, 1392, 1964.
22. **Gibaldi, M. and Perrier, D.**, *Pharmacokinetics,* 2nd, Marcel Dekker, New York, 1982, chap. 11.
23. **Snedecor, G. W. and Cochran, W. G.**, *Statistical Methods,* 7th ed., The Iowa State Univeristy Press, Ames, Iowa, 1980, chap. 6.
24. PC-SAS, SAS Institute, Cary, NC.
25. **Snedecor, G. W. and Cochran, W. G.**, *Statistical Methods,* 7th ed., The Iowa State University Press, Ames, Iowa, 1980, chap. 12.
26. **Snedecor, G. W. and Cochran, W. G.**, *Statistical Methods,* 7th ed., The Iowa State University Press, Ames, Iowa, 1980, chap. 13.
27. **Snedecor, G. W. and Cochran, W. G.**, *Statistical Methods,* 7th ed., The Iowa State University Press, Ames, Iowa, 1980, chap. 8.
28. **Winer, B. J.**, *Statistical Principles in Experimental Design,* 2nd, McGraw-Hill, New York, New York, 1971.

Chapter 10

SPECIALIZED TECHNIQUES — CONGENITALLY ATHYMIC (NUDE) ANIMAL MODELS

George J. Klain and Kenneth E. Black

TABLE OF CONTENTS

I. INTRODUCTION

The nude mouse mutant was first observed by N. R. Grist in 1962 in inbred stock of albino mice in the Virus Laboratory, Ruchill Hospital, Glasgow.[1] The mutant was described in detail by Flanagan in 1966 who gave the name "nude", symbol nu, to the mutant.[2] Flanagan also found that this mutation was inherited as autosomal recessive, and from intercross matings of heterozygous nu/+ mice, 25% were nude. A characteristic feature of the mutant is a nearly complete lack of body hair. Sparse hair growth may appear occasionally. Nude mice can be identified at birth by the absence of whiskers or by poorly developed crinkled whiskers.[3] At birth the body weight of nude mice are comparable to those of phenotypically normal littermates, but nude mice gain progressively less weight. In 1968 it was reported that the homozygous nude mouse lacks a thymus.[4] Since the nude mouse is immunodeficient it has been an ideal animal model in studies concerned with some congenital and acquired immunodeficiency syndromes and infections, the thymus humoral factor, and human malignant tumors. In addition, it has been established that nude mice readily accept skin grafts from a wide variety of mammals, including man, and from birds. The grafts maintain all the morphological and ultrastructural features associated with normal skin,[5,6] maintain their own functional properties and are metabolically active.[7] Human skin-grafted nude mice have been used successfully in percutaneous absorption studies.[7-9] The use of human skin grafted nude mice also allows the investigator: (1) to study dermal metabolism of toxic compounds in viable human skin without risk to humans and (2) to compare metabolic transformation of such compounds in the skin grafted to the mouse. However, identification of a metabolite in the blood following a topical application of a chemical does not necessarily provide a clue to its site of origin. The metabolite may have been formed at a site far removed from the application skin site, in the skin or in both. In addition, degradative pathways for a specific compound in the skin may be different from those in other organs and tissues. A recent development of a congenitally nude (athymic) rat/human skin flap model[10,11] permits not only a direct assessment of absorption across skin, but also a direct assessment of dermal metabolic transformations of topically applied test compounds. The skin flap consists of a split thickness (human, pig) grafted to the subcutaneous surface of the epigastric skin of the abdomen of the athymic nude rat. The sandwich flap is then isolated with its supplying vasculature, transferred to the dorsum of the rat through a subcutaneous tunnel, and sutured in place.

In contrast to athymic mice or rats, hairless mice and hairless rats have functional thymus and will not accept skin grafts unless their immune system is suppressed. Yet these two animal models are frequently used in *in vivo* skin absorption studies since they do not require pretreatments, such as shaving or depilitation prior to experimentation. These pretreatments may affect the integrity of the stratum corneum.

II. PREPARATION OF MODELS

A. HUMAN SKIN GRAFTED NUDE MOUSE

1. Housing for Animals

Since nude mice are immunodeficient they are more susceptible to infection than mice with a normal immune system. Thus, good management practices in the animal facility are essential for the maintenance of high quality and healthy animals, in particular of those recovering from skin grafting procedures. After receipt, the mice should be kept in quarantine for 2 weeks in limited-access rooms and held singly in plastic cages with a sealed air filter. Individual cages are kept in germfree isolators which supply filtered air. The cages and the isolators are available from Hazelton Systems Inc., Aberdeen, MD. The mice must be kept isolated from other laboratory animals, and laboratory personnel as well. Attending personnel must maintain a very high level of cleanliness, must wear caps, masks, sterile clothing, gloves and shoe covers, and should not handle other rodents. Eating, drinking, or smoking must not be allowed in the room. Before use the cages, cage tops and filters, water bottles, stoppers, and drinking tubes must be autoclaved. Commercial mouse diet and cage bedding must also be sterilized. It is important that all mice have an adequate supply of acidified water (0.1 ml of 6 N hydrochloric acid per liter of water). Nude mice drink more water than normal mice, apparently because of the higher loss of moisture from the bare skin. It is recommended that mice be transferred to clean cages weekly. Since nude mice lose body heat more rapidly than normal mice the room temperature should be kept around 25°C and the relative humidity around 50% to minimize the loss of moisture from the skin. A uniform dark-light cycle of 12 h each is recommended. Nude mice are highly temperamental. Fighting mice must be separated to prevent undue skin and other injuries.

The authors have always strictly followed proper husbandry practices to maintain healthy and vigorous nude mice for skin grafting and during the 6-week post grafting period. However, on several occasions healed skin grafted mice were removed from the isolation room and kept in the filter-topped plastic cages in the authors' laboratory for up to 13 months. Our technical personnel had free access to the laboratory but no rodents were housed or used in the same laboratory. No wasting of these animals or other ill effects were observed. The mice were briefly handled only while the cages were being cleaned. Our observations may indicate that the trauma associated with skin grafting procedures is one of the significant factors affecting the recovery of the nude mouse after operation. Following recovery, they could be kept in ordinary facilities, as long as they are separated from other rodents.

2. Grafting Techniques

Only healthy mice, at least 6 weeks of age or older should be used for skin grafting. All experimental procedures must be done under a biological hood in a clean room. The mice are anesthetized by an intraperitoneal injection of chloral hydrate (400 mg/kg) or sodium pentabarbital (60 mg/kg). The site of the graft may vary, but the thorax or the back are the most common sites, since it is difficult for the mouse to reach these areas and disturb the graft. A circular piece of the mouse skin about 1 to 2 cm in diameter is excised with sterile instruments, without disturbing the underlying vascularized connective tissue. The prepared skin graft disc is placed into the graft bed and dressed with a surgical tape (Micropore, 3M, St. Paul, MN). The tape can be removed in 2 to 3 weeks. Grafted mice are placed in individual filter-topped cages to keep other mice from disturbing the graft. At about 6 weeks the grafted mouse is ready for experimentation (Plate 1*). Two recent publications describe the grafting techniques in detail.[12,13]

The thickness of the transplanted skin is an essential component of the successful grafting technique. Full thickness grafts usually do not succeed unless the donor skin is thin, for example from an eyelid. The donor skin must be dermatomed to a split thickness sheet of skin of desired thickness (0.2 to 0.7 mm), a range which corresponds in thickness to the depth of the mouse graft bed. A sterile cork borer is used to cut the dermatomed skin into discs, usually 1 to 2 cm in diameter. The thickness of the disc determines what components of the skin remain functional in the graft. Skin discs 0.1 mm thick or less usually do not take. The discs should contain some papillary dermis to be successful. Thicker discs from an area containing sebaceous glands will produce grafts with functional glands, and deeper grafts will retain hair bulbs. In our Institute percutaneous penetration studies have been conducted with 0.7 mm thick pig or human skin grafts.[8] On one occasion long pig bristles were observed on the graft of a mouse kept over a year in the laboratory (Plate 2*). We have not determined the time required for the hair bulb to become active after grafting. The

---

*   Plates 1 and 2 follow p. 184.

activity of hair bulbs and other adnexal structures in skin grafts has not been studied adequately. Skin grafts frequently thicken during the months following emplacement; the effect of this thickening on percutaneous absorption also has not been determined.

B. HUMAN SKIN GRAFTED NUDE RAT FLAP MODEL

The nude mutation of rat was first observed at the Rowett Research Institute, UK in 1953, and a breeding colony was established in 1975. The mutant was described in 1978.[14] A similar athymic nude rat mutant appeared in New Zealand.[15] In many respects the characteristics of the nude mouse and the nude rat are similar. Their immune system is depressed,[16,17] and they may be partially or completely hairless. Depending on the objective of a specific study, the nude rat could be a suitable alternative to the nude mouse. Since the rat is larger than the mouse, it could accept larger grafts. The rat provides a larger volume of urine and can be bled more frequently. Unfortunately, the nude rat usually rejects split thickness human skin grafts over a 4-week time period.[18] A low dose of cyclosporine (25 mg/kg/d for 3 weeks, Sandoz, East Hanover, NJ) injected subcutaneously, seems to overcome the rejection process.

1. Housing

Athymic nude rats should be housed in clean, limited-access rooms. The rats can be kept singly in filter-topped plastic cages in moveable units which provide filtered air. Food, water, bedding and cages should be sterilized to prevent introduction of infection.

2. Preparation of Human Skin Grafted Flap

Donor human skin used in the preparation of the flap is dermatomed to a thickness of 0.5 to 0.7 mm. The skin may be used immediately or preserved at 4°C in a tissue culture medium, such as RPMI-160 with 10% bovine serum (Gibco Laboratories, Grand Island, NY), and used within 72 h.[11] The preparation of the flap is accomplished in several distinct steps. In the initial step the rat is anesthetized (Ketamine, 100 mg/kg, injected i.p.), and a skin sandwich is formed as a flap by grafting a split thickness skin graft to the subcutaneous surface of the skin in the area supplied by the superficial epigastric vessels.[10] During the next 2 weeks blood vessels of the rat skin form a vascular network in the donor skin. The sandwich flap is then detached from the surrounding abdominal skin and the resulting skin defect is corrected with a corresponding piece of split thickness rat skin. The superficial epigastric artery and vein are isolated in the subsequent step, utilizing appropriate micro-surgical techniques.[19] The isolated skin sandwich flap, together with a translocated rat abdominal vasculature, is transferred by a subcutaneous route to the dorsum of the rat, and sutured to the dorsal skin. Another two weeks will be required to complete the healing process (Figure 1). A protective wrap, consisting of stretch bandage (Conform, Kendall, Boston, MA) and surgical tape (Micropore, 3M, St. Paul, MN) is applied around the torso of the rat to prevent it from disturbing the flap prior to experimentation. During this recovery period the rats are individually caged with filtered air and indirect lighting. Food, water, and bedding are autoclaved. Drinking water containing cyclosporine (0.07 mg/ml) is acidified (pH = 2.5), and the indirect lighting retards the deterioration of cyclosporine. The water is changed every two days. Additional surgical and technical details can be found in the original publications.[10,11]

III. STUDIES WITH HUMAN SKIN GRAFTED NUDE MOUSE

The human skin grafted nude mouse model has been of value in numerous studies concerned with various facets of skin biology and pathology, including percutaneous penetration and metabolism of chemical compounds. Although the model has been validated,

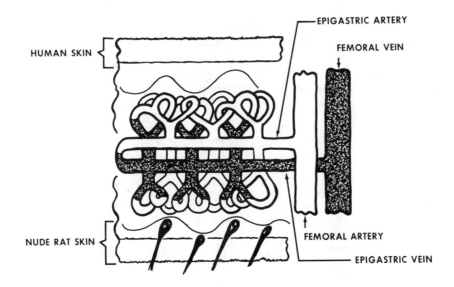

FIGURE 1. Schematic of the human skin grafted nude rat flap

and found to have a substantial advantage over some other animal models, its use is limited by the goal of a specific study.

A. ADVANTAGES AND LIMITATIONS OF MODEL

1. Skin grafts maintain their structural, morphological and functional integrity for prolonged periods of time
2. The grafting procedure is simple to perform and does not require any expensive instrumentation
3. Percutaneous penetration/metabolism of toxic or radioactive chemicals in viable skin can be studied without risk to humans
4. Metabolic transformations in the skin grafts can be compared with those in other tissues and organs of the grafted mouse
5. No drugs are required for the maintenance of the graft
6. Penetration properties of human skin grafted onto athymic mice and ungrafted human skin are similar

The model may have the following limitations:

1. The size of the animal limits the frequency of blood sampling in time studies
2. The role of the missing portion of the dermis in split thickness skin grafts cannot be assessed in dermal penetration or metabolic studies, particularly with highly lipophilic compounds
3. The cost of maintaining the colony
4. The requirement for clean and limited-access rooms

B. DISTRIBUTION AND FATE OF TOPICALLY APPLIED COMPOUNDS IN THE SKIN GRAFT

Mice are killed by cervical dislocation and the skin graft is cleansed with a cotton swab lightly soaked with ethanol. The graft is quickly excised, weighed, and chilled in ice-cold saline. The graft is sliced parallel to skin surface, using a microtome (American Optical Corp., Buffalo, NY) to prepare epidermal and dermal slices of desired thickness (40 to 50 μm). Individual tissue slices are combusted in a tissue oxidizer (Packard Instruments, Des

TABLE 1
Solvents Used for the Extraction of Test
Compounds

Solvent	Dielectric constant (20°C)	Polarity
n-Hexane	1.890	Least polar
Heptane	1.924	
Cyclohexane	2.023	
Carbon tetrachloride	2.238	
Benzene	2.284	
Diethylether	4.335	
Chloroform	4.806	
Ethyl acetate	6.02[a]	
Dichloromethane	9.08	
Pyridine	12.3[a]	
n-Butanol	17.8	
n-Propanol	20.1[a]	
Acetone	20.7[a]	
Ethanol	24.30[a]	
Methanol	32.63	
Water	80.37	Most polar

[a]   At 25°C.

Plaines, IL, Harvey Industries, Hillsdale, NJ). Radioactivity in the resulting oxidation products is determined in a liquid scintillation spectrometer (Packard Instruments, Des Plaines, IL). Alternatively, tissue slices are dissolved in 2 ml of Hyamine solution (Packard Instruments, Des Plaines, IL) by heating at 60° C for 2 h. The solutions are diluted with aqueous counting solution and the radioactivity is determined in a liquid scintillation spectrometer. A profile of radioactivity in the graft can be constructed by plotting DPMs versus the corresponding tissue slices. Using this technique, it has been recently reported that the lower portion of the dermis obtained from human skin grafts treated with topical ^{14}C-labeled alpha-tocopherol contains more radioactivity than the upper portion of the dermis plus the epidermis.[20] Significant dermal accumulations were reported with pig skin after topical application of parathion.[21] These findings suggest that the dermis may act as a barrier or reservoir for highly lipophilic compounds.

Metabolic activity in human skin grafts was demonstrated by studying the incorporation of ^{14}C from topical ethanolamine into various nitrogenous components of the skin.[7] The grafts were homogenized in ethanol-ethyl ether and the lipid components were extracted with petroleum ether. The lipid-free residue was hydrolyzed with boiling 6 N HCl for 20 h and the radioactive components were separated by ion-exchange column chromatography, and identified by co-chromatography and colorimetric tests, using authentic compounds.

In percutaneous penetration studies, using an unlabeled compound, the compound must be initially extracted into an organic solvent. Highly lipophilic compounds in an aqueous environment will readily dissolve in a nonpolar solvent (i.e., hexane). Other compounds containing, for example, hydroxyl groups will require more polar solvents for aqueous extraction. Table 1 lists selected solvents, in an increasing order of polarity, commonly used to extract organic compounds from tissues or biological fluids.[22] The dielectric constants are also included, as they may be taken as indicators of polarity. The pH of the aqueous phase should be adjusted to ensure that the compound is not in the ionized form and can be extracted by organic solvents. The meaningful use of an unlabeled test compound depends on the sensitivity of analytical techniques to measure the concentration in the graft. The concentration may be low due to a rapid penetration. In addition, it may be difficult to completely extract the compound from the graft. Appropriate analytical techniques to quantify the compound in the skin slices include high performance liquid chromatography, gas-liquid chromatography with various detectors, gas chromatography-mass spectrometry, various bioassays, enzyme immunoassays, spectrophotometric, and spectrofluorometric methods. These are techniques of general acceptibility. Their use will depend on the design of each study.

C. MASS BALANCE STUDIES FOR MEASUREMENT OF PERCUTANEOUS ABSORPTION

After application of the radioactive test compound, mice are killed at predetermined time intervals, and blood, organs and various tissues, skin grafts, and remaining carcass are assayed for radioactivity. Expired carbon dioxide, urine, and feces are also collected for radioactivity determination. The purpose of this procedure is to account for all of the absorbed radioactivity, and to express the results as a fraction of the applied radioactive dose. Mice are a convenient animal model for mass balance studies because of a small body mass. The tissues and organs are of a convenient size for storage or experimental procedures.

1. Collection of Biological Samples

Metabolism cages allow quantitative separate collection of urine, feces, expired carbon dioxide, and other volatile metabolites. The animal has a free access to food and water. Metabolism cages for small laboratory rodents are usually constructed from Pyrex glass or from various polymeric materials. Urine and feces are collected daily to determine the completeness of excretion of the test compound. After removal of the fecal material, the collection area is rinsed with distilled water and ethanol to ensure a complete recovery of the urine. The volume of urine plus the washings are recorded and kept in a refrigerator. The fecal material is kept frozen until analyzed. Air is drawn through the metabolism cage and through a 2% solution to NaOH[7] or an organic base, such as ethanolamine, or Hyamine® to absorb expired $^{14}CO_2$. Other voltatile metabolites in expired air may be retained on a glass column containing Tenax® absorbent.

At the end of the collection period, mice are killed by cervical dislocation, and the skin graft and other tissues are excised, cooled in ice-cold saline and weighed.

2. Determination of Radioactivity

Aliquots of urine, sodium hydroxide, or organic base solutions are pipetted into scintillation vials and mixed with 10 ml of aqueous counting solution (ACS). Tenax® samples are placed directly into scintillation vials and the columns are flushed with 10 ml ACS. Radioactivity in the solutions is determined in a liquid scintillation spectrometer equipped with an automatic external standard.

Fecal samples are freeze-dried and thoroughly homogenized. Skin grafts are cut into small (50 mg) pieces, and carcass, tissues, and organs are homogenized with an equal volume of water. Radioactivity in tissue preparations (equivalent to about 150 mg tissue) is then determined in a tissue oxidizer.

IV. STUDIES WITH THE HUMAN SKIN GRAFTED NUDE RAT FLAP MODEL

The human skin grafted nude rat flap model is the latest animal model developed in an attempt to gain a further insight into the mechanisms of transdermal absorption across human skin.

A. ADVANTAGES AND LIMITATIONS OF MODEL

1.   Since the flap has an independent but accessible vascular supply, transdermal flux of the test compound can be readily determined

2.   The dermal degradation of the test compound can be assessed before it reaches the systemic circulation
3.   The flap can be used in another experiment after a 2 to 3 week interval between experiments
4.   The flap allows a comparison of the flux between the human and the rat skin components
5.   Effects of external factors (temperature, humidity, absorption enhancers, drugs) on transdermal flux can be performed

The model has the following limitations:

1.   Preparation of flap requires specialized techniques and skill in microsurgery
2.   The model requires low-dose oral cyclosporine which can affect penetration[18]
3.   Maintenance of the model is labor intensive in a separate animal room

B. MEASUREMENT OF PERCUTANEOUS ABSORPTION

Animals are anesthetized with ketamine (Ketajet 100 mg/ml, Parke Davis, Morris Plains, NJ) 10 mg/100 g body weight, and the animal is placed on a thermostatically controlled glass water bed (Laboratory Glass Apparatus, Berkeley, CA) to maintain normal body temperature. To prevent dehydration, 2.5 ml of normal saline is injected i.p. Vessels supplying and draining the flap are exposed through an incision over the inguinal ligament. Further dissection with microsurgical instruments (Accurate Surgical and Scientific Instrument Corp., Westbury, NY) is made to expose the femoral vein. This femoral vein is the site of sampling of blood draining the flap. A circular suture on the ventral wall of the femoral vein is made to mark the site of venotomy and for hemostasis in case of excessive bleeding. This procedure preserves the vascular drainage of the flap in case of profuse bleeding and could close the venotomy site with lesser chance of causing vascular stricture. A venotomy is also made in the contralateral femoral vein to sample systemic blood.[10,23] Mean blood levels of radioactivity in the femoral vein draining the flap may then be plotted along with the corresponding levels in the contralateral femoral vein representing systemic levels of radioactivity. The recoveries of radioactivity in the urine, along with the data obtained after subcutaneous administration of the test compound, allow the investigator to calculate the percutaneous penetration.

C. ANALYSIS OF BLOOD FOLLOWING TOPICAL APPLICATION

Aliquots of blood (10 to 20 μl) are placed directly into oxidation cups and the radioactivity is determined in a tissue oxidizer. Alternatively, blood aliquots are placed into scintillation counting vials, and 50 to 100 μl of 30% hydrogen peroxide (Fisher Scientific, Fairlawn, NJ) are added to decolorize the samples. After standing for 4 to 6 h, 10 ml of counting solution (Optifluor, Packard Instrument Co., Downers Grove, IL) is added. The radioactivity in the vials is determined in a scintillation spectrometer.

If an unlabeled test compound is used in this study, the red blood cells are separated from the plasma by centrifugation (2500 RPM/10 min) at 5°C. Any dilution of the whole blood should be done with physiological saline to prevent rupture of the cells. The initial procedure in preparation of plasma samples for analysis is to precipitate plasma proteins. The test compound can then be extracted from the protein-free filtrate with an organic solvent. Commonly used acidic compounds for protein precipitation are perchloric acid, trichloroacetic acid and tungstic acid. These reagents, however, could have an adverse effect on the test compound, and should be tested before experimentation. Alternatively, methanol or ethanol can be used to denature and precipitate plasma proteins. Three volumes of ethanol are usually used to precipitate all plasma proteins. Another convenient solvent for the precipitation of plasma proteins and the extraction of lipophilic compounds is acetonitrile.[24] Plasma is mixed with an equal volume of acetonitrile, and to this is added a mixture of sodium bisulfate, H_2O and sodium chloride (1:4) in the amount of about 200 mg/ml. The solution is mixed, centrifuged, and the upper phase is taken for analysis by an appropriate procedure.

APPENDIX

A. Commercial source of nude mice: 1. Harlan Industries, Inc., P.O. Box 29176, Indianapolis, IN 46229; 2. Simonsen Laboratories, Inc., 1180 C Day Road, Gilroy, CA 95020; 3. Bantin and Kingman Co., 3403/3421 Yale Way, Fremont, CA 94538.

B. Commercial sources of nude rats: 1. Harlan Industries, Inc., P.O. Box 29176, Indianapolis, IN 46229; 2. Sprague-Dawley Animal Facility, Indianapolis, IN; 3. Small Animal Breeding Facility, National Institute of Health, Bethesda, MD.

REFERENCES

1. **Isaacson, J. H. and Cattanach, B. M.,** Report, *Mouse News Lett.,* 27, 31, 1962.
2. **Flanagan, S. P.,** "Nude", a new hairless gene with pleiotropic effects in the mouse, *Genet. Res. (Camb.),* 8, 295, 1966.
3. **Rygaard, J. and Friis, C. W.,** The husbandry of mice with congenital absence of the thymus (nude mice), *Zeit. Versuchstierkd.,* 16, 1, 1974.
4. **Pantelouris, E. M.,** Absence of thymus in a mouse mutant, *Nature (London),* 217, 370, 1968.
5. **Manning, D. D., Reed, N. D., and Shaffer, C. G.,** Maintenance of skin xenographs of widely divergent phylogenetic origin on congenitally athymic (nude) mice, *J. Exp. Med.,* 138, 488, 1973.
6. **Reed, N. D. and Manning, D. D.,** Long-term maintenance of normal human skin on congenitally athymic (nude) mice (37318), *Proc. Soc. Exp. Biol. Med.,* 143, 350, 1973.
7. **Klain, G. J., Reifenrath, W. G., and Black, K. E.,** Distribution and metabolism of topically applied ethanolamine, *Fund. Appl. Toxicol.,* 5, S127, 1985.
8. **Reifenrath, W. G., Chellquist, E. M., Shipwash, E. A., Jederberg, W. W., and Krueger, G. G.,** *Br. J. Dermatol.,* 111, Suppl. 27, 123, 1984.
9. **Scott, R. C. and Rhodes, C.,** The permeability of grafted human transplant skin in athymic mice, *J. Pharm. Pharmacol.,* 40, 128, 1988.
10. **Krueger, G. G., Wojciechowski, Z. J., Burton, S. A., Gilhar, A., Heuther, S. E., Leonard, L. G., Rohr, U. D., Petelenz, T. J., Higuchi, W. I., and Pershing, L. K.,** The development of a rat/human skin flap served by a defined and accessible vasculature on a congenitally athymic (nude) rat, *Fund. Appl. Toxicol.,* 5, S112, 1985.
11. **Wojciechowski, Z., Pershing, L. K., Huether, S., Leonard, L., Burton, S. A., Higuch, W. I., and Krueger, G. G.,** An experimental skin sandwich flap on an independent vascular supply for the study of percutaneous absorption, *J. Invest. Dermatol.,* 88, 439, 1987.
12. **Black, K. E. and Jederberg, W. W.,** Athymic nude mice and human skin grafting, in *Models in Dermatology,* Vol. 1, Maibach, H. I. and Lowe, N. J., Eds., Karger, Basel, 1985, 228.
13. **Briggaman, R. A.,** Human skin graft-nude mouse model: techniques and application, in *Methods in Skin Research,* Skerrow, D. and Skerrow, C. J., Eds., John Wiley & Sons, 1985, chap. 9.
14. **Festing, M. F. W., May, D., Connors, T. A., Lovell, D., and Sparrow, S.,** An athymic nude mutation in the rat, *Nature (London),* 274, 365, 1978.
15. **Berridge, M. V., O'Kech, N., McNeilage, L. J., Heslop, B. F., and Moore, R.,** Rat mutant (NZNU) showing "nude" characteristics, *Transplantation,* 27, 410, 1979.
16. **Brooks, C. G., Webb, P. J., Robins, R. A., Baldwin, R. W., and Festing, M. R. W.,** Studies on the immunobiology of rnu/rnu "nude" rats with congenital aplasia of the thymus, *Eur. J. Immunol.,* 10, 58, 1980.

17. **Vos, J. G., Kreeftenberg, J. G., Kruijt, B. C., Kruizinga, W., and Steerenberg, P. A.,** The athymic nude rat. II. Immunologic characteristics, *Clin. Immunol. Immunopathol.,* 15, 229, 1980.
18. **Pershing, L. K. and Krueger, G. G.,** New animal models for bioavailability studies, in *Pharmacology and the Skin,* Vol. 1, Shroot, B. and Schaefer, H., Eds., Karger, Basel, 1987, 57.
19. **Buncke, H. J., Chater, N. L., and Szabo, Z.,** *The Manual of Microvascular Surgery,* Ralph K. Davies Medical Center, Microsurgical Unit, San Francisco, 1975.
20. **Klain, G. J. and McDermott, B.,** Absorption and tissue distribution of ^{14}C-labeled alpha tocopherol (Vit E) in human skin grafted athymic nude mice, *The Japanese - United States Congress of Pharmaceutical Sciences,* (Abstr.), S78, Honolulu, December 2 to 7, 1987.
21. **Reifenrath, W. G., Kurtz, M. S., and Hawkins, G. S.,** Properties of the dermis during in vitro and in vivo percutaneous absorption measurements, *The Japanese - United States Congress of Pharmaceutical Sciences,* (Abstr.), S293, Honolulu, December 2 to 7, 1987.
22. **Trappe, W.,** Die Trennung von biologischen Fettstoffen aus ihren naturlichen Gemischen durch Anwendung von Adsorptionssaulen. Die eluotrope Reihe der Losungsmittel, *Bioch. Zeit.,* 305, 150, 1940.
23. **Bernardo, E. S. and Reifenrath, W. G.,** personal communication, 1988.
24. **Mathies, J. C. and Austin, M. A.,** Modified acetonitrile protein-precipitation method of sample preparation for drug assay by liquid chromatography, *Clin. Chem.,* 26, 1760, 1980.

Chapter 11

SPECIALIZED TECHNIQUE: THE ISOLATED PERFUSED PORCINE SKIN FLAP (IPPSF)

Nancy Ann Monteiro-Riviere

TABLE OF CONTENTS

I. INTRODUCTION

The assessment of the percutaneous absorption of drugs and xenobiotics has been limited by the relatively simple nature of previous experimental models. The study of the function of skin in toxicology and dermatology would be facilitated if *in vitro* models possessed a viable and defined venous and arterial vascular supply. These limitations in experimental methodology provide the incentive for developing an alternative animal model with an intact microcirculation. It is necessary to select a suitable species in which the skin is structurally and physiologically comparable to that of humans.[1-2] The surgical manipulations required to create a skin preparation with an intact microcirculation preclude the use of fresh human skin because of ethical considerations. The aim of this chapter is to fully describe the development and utilization of a unique *in vitro* perfused skin model, the isolated perfused porcine skin flap (IPPSF).

Many structural differences exist between the skin of domestic and laboratory animals. These include the arrangement of hair follicles and density of hair. Differences also exist in the thickness of the epidermis and dermis between species and within the same species in various regions of the body.[3] However, the basic architecture of the integument is similar in all mammals. Numerous authors have alluded to the similarity of pig and human skin based on structure and function. Swine resembles human skin in having a sparse hair coat, a relatively thick epidermis, similar epidermal turnover kinetics, lipid composition, enzyme histochemistry, and carbohydrate biochemistry. The dermal microcirculation is reported to be similar to man as is the arrangement of dermal collagen and elastic fibers.[4-8]

Topical application and absorption of the xenobiotics *in vivo* is characterized by measuring the absorbed agent and/or its metabolites in blood or in excretory products. There are numerous *in vitro* methods whereby penetration and absorption of xenobiotics are measured through cadaver skin placed in diffusion cells with either static or flowing receptor fluid. The previous chapters in this text should be consulted for a full description of these *in vivo* and *in vitro* techniques. Often, *in vivo* studies are confounded by systemic metabolism of the compound or by sequestration of the agent into body fat, and classical *in vitro* studies are not optimal if the agent is extremely lipid soluble or metabolized by the skin.[9-10] The interaction of the cutaneous microcirculation of the absorption profile or cutaneous metabolism of the compound cannot be studied in these models.

The development of a viable *in vitro* perfused skin flap with intact cutaneous microcirculation would allow optimal percutaneous absorption studies to be conducted. By utilizing an isolated perfused preparation, experimental conditions could be controlled and the confounding influence of systemic processes (i.e., metabolism) would be eliminated. Also rigorous pharmacokinetic modeling could be performed because arterial, venous and tissue samples are readily available for collection of data. An ideal preparation should be perfused in a controlled environment where temperature and humidity could be varied. In order to accomplish this latter objective, a tubed skin flap is required so that viable dermis is not exposed to the environment. If this is not accomplished, the skin preparation would have to be maintained in a 100% humidified oxygenated atmosphere. Finally, the preparation should be of a suitable size and possess a surface texture which is comparable to that seen in man so that application of transdermal delivery systems intended for human use could be directly tested in the model. A single-pedicle, axial-pattern, tubed skin flap would meet these requirements. Pigs possess a direct cutaneous vasculature in the ventral abdomen which is amenable to surgical production of such a skin flap. This attribute coupled with the previously mentioned structural and functional similarity of porcine and human skin were the primary reasons that an isolated perfused porcine skin flap model was developed.

II. DEVELOPMENT OF SURGICAL PROCEDURE

A two-stage surgical procedure is utilized to create the IPPSF. Stage I involves creating the preparation and Stage II its subsequent harvest for isolated perfusion. The discussion below describes the creation of a single flap based on either the right or left superficial epigastric artery. However, two flaps can be simultaneously created on a single pig in a surgical procedure because of the animal's symmetry. This procedure decreases interflap variability, provides for a paired control preparation, and minimizes cost and time within an experiment.

A. GROSS ANATOMY

Female Yorkshire pigs weighing approximately 40 kg were first anesthetized, heparinized, exsanguinated, and embalmed with red (arterial) and blue (venous) injectable latex to dissect and investigate the cutaneous vasculature in the caudolateral abdominal region. The caudal superficial epigastric artery and its paired venae comitantes were found to be the major vascular supply to this caudolateral abdominal region. Cranial to the caudal-most teat, and at the caudal extent of the cutaneous trunci muscle, the abdominal skin was primarily supplied by musculocutaneous vasculature. Therefore, a single-pedicle, axial-pattern, tubed flap was created around the caudal superficial epigastric artery.

B. STAGE I SURGICAL PROCEDURE

Pigs were immobilized with atropine sulfate (0.04 mg/kg i.m.) and xylazine hydrochloride (0.2 mg/kg i.m.). For anesthesia, ketamine hydrochloride (11 mg/kg i.m.) was used while the pig remained in a dorsal recumbent position and halothane was delivered by endotracheal tube. Pigs were prepared for routine aseptic surgery in the caudal abdominal region, and an area 4 cm × 12 cm just caudolateral to the umbilicus was demarcated with a sterile marking pen. A scalpel was used to cut through the skin and the subcutaneous tissue, and then the skin flap was lifted and pulled back without damaging the underlying vascular supply (Figure 1). The flap was then tubed and closed with size 3-O-polypropylene suture in a simple continuous pattern, and excess fat was trimmed. The tubed flap was secured to the surrounding abdominal skin via one or two sutures and bandaged. The healing of the axial-pattern tubed flaps were evaluated by visual inspection along with fluorescein angiography. The skin flaps survived well during the interval before Stage II harvest, if the animals were housed in spacious pens and given suitable distractions (i.e., toys) to prevent self-mutilation. Extensive bandaging is not recommended because of the tendency to compromise the skin flap's circulation.

C. ASSESSMENT OF THE MICROCIRCULATION
1. Fluorescein Angiography

Anesthetized pigs were prepared for aseptic surgery and skin flaps were raised as described above. In these initial studies, the length and width of the skin flap were varied to determine the optimal size. Five milliliters of 10% fluorescein (Fluorescite Injection®, Alcon Lab. Inc., Ft. Worth, TX) was infused into the flap to help demarcate areas that were perfused or nonperfused. These studies were repeated at various times after healing to help determine surviving length of the flap. From these studies, a skin flap 12 cm long and 4 cm wide was judged optimal.

2. Fluoroscopy and Contact Radiography

The skin flaps were then subjected to contact radiography to compare and determine the architecture of vessels and extent of the surviving length of the flap. Through incision and blunt dissection, the femoral artery and the caudal superficial epigastric arteries were exposed in additional pigs. These arteries were then cannulated to allow infusion of 60% meglumine iothalamate (Conray®, Mallinckrodt Inc., St. Louis, MO) for *in vivo* angiography. These fluoroscopy studies demonstrated that the superficial epigastric artery supplied

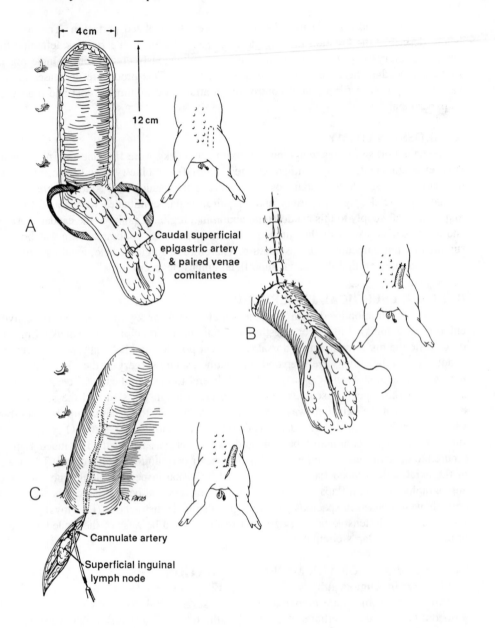

FIGURE 1.   Surgical procedures for preparation of *in vitro* isolated perfused porcine skin flaps. (A) Stage one procedure: single pedicle, axial-pattern tubed flap supplied by the caudal superficial epigastric artery and its paired venae comitantes. (B) The donor site and tubed flap are closed and secured to the surrounding abdominal skin via one or two sutures and bandaged. (C) Stage two procedure: 2 d later, the tubed flap is harvested following cannulation of the caudal superficial artery and one of the venae comitantes and transferred to the isolated organ perfusion apparatus. (Reprinted from Riviere et al.[12]).

the skin flap in a living animal and venous return occurred via the associated paired venae comitantes.

Pigs were heparinized, the superficial epigastric artery cannulated, the skin flap removed, small vessels cauterized, and micropulverized 30% barium sulfate (Micropaque® Powder, Picker Corp., Highland Heights, OH) 4% gelatin solution (Fisher) was infused to visualize

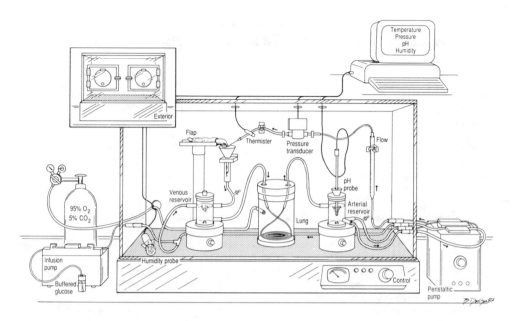

FIGURE 2. Schematic diagram of the perfusion apparatus used to maintain the IPPSF at constant temperature, humidity, pressure and perfusate flow. (Reprinted from Carver et al.[19])

the cutaneous microcirculation. These studies demonstrated that superficial and deep cutaneous capillary beds were being perfused in the skin flap. In additional studies, formalin-fixed, barium-perfused flaps were subjected to microangiography study. As shown in Plate 3*, both the superficial and deep capillary beds were filled with barium.

D. STAGE II SURGICAL PROCEDURE

Two days later the axial pattern tubed flap is harvested (Figure 1). Pigs are preanesthetized, induced and maintained on anesthesia as described above. Fluorescein angiography was performed to ascertain that the skin flap had survived in its entire length. After aseptic surgery, the caudal superficial epigastric artery was cannulated with PE 50 polyethylene (I.D. 58 mm, O.D. 97 mm, Fisher) tubing secured by a few stay sutures. The flap is then heparinized and flushed with 20 ml of saline to clear the flap of blood and to establish patency of the vessels.

The tubed flap is then immediately transferred to the isolated perfusion chamber. The pig is recovered from anesthesia and the wound is left open to heal by granulation and is then returned to its prior habitat. The stage I surgical procedure takes a trained technician 90 to 120 min while a stage II procedure takes 15 min of actual surgical time. For a more detailed description of the surgical procedure see Bowman et al.[11]

III. ISOLATED PERFUSION

A. CHAMBER DESIGN

The perfusion apparatus (Figure 2) (Riviere Isolated Perfused Skin Flap Chamber, Diamond Research, Raleigh, NC) is a recirculating/nonrecirculating system regulated for the relatively low perfusate flow rates (0.3 to 5.5 ml/min) seen in skin flaps. Flaps are perfused with a mean arterial pressure ranging from 30 to 70 mmHg. A humidified (%RH = 40 to

* Plate 3 follows p. 184.

TABLE 1

Preparation of Isolated Perfused Porcine Skin Flap (IPPSF) Media

Ingredients
  NaCl, 13.78 g
  KCl, 0.71 g
  $CaCl_2$, 0.56 g
  KH_2PO_4, 0.32 g
  $MgSO_4 \cdot 7H_2O$, 0.58 g
  $NaHCO_3$, 5.50 g
  Dextrose, 2.40 g
  Bovine serum albumin, fraction V, 90.0 g (I.C.N. Biochemicals, Cleveland, OH)
  Glass distilled water, q.s.
Methods
  Add approximately 1700 ml of glass distilled water and a stir bar to a 4000 ml beaker
  Add ingredients 1 to 7 to the beaker and mix well, adjusting pH to approximately 7.4
  Add the ingredient BSA (#8) only after step 2. Add BSA slowly to the beaker while stirring until all of it has
    gone into solution. It may take an hour or so for all BSA to dissolve
  Add 10 ml of sodium heparin (Lyphomed Inc., Rosemont, IL) (1000 USP units/ml)
  Add 0.1 ml of penicillin G sodium (Squibb and Sons, Inc., Princeton, N.J.) (250,000 u/ml). (Note that this
    drug degrades when in solution greater than 7 days.)
  Add 0.4 ml of amphotericin B (Squibb and Sons, Inc., Princeton, N.J.) (5.0 ml/ml). (Expires in 7 days.)
  Add 0.25 ml of amikacin (Fort Dodge Lab. Inc, Fort Dodge, IA) (250 mg/ml).
  Bring volume up close to 2000 ml. Mix well, balance pH to 7.45 ± 0.03 with NaOH and/or HCl.
  Remove stir bar; pour solution into a 2000 ml volumetric flask
  Bring volume to 2000 ml with glass distilled water

100) plexiglass chamber encloses this apparatus and can be maintained from 25 to 44°C. A typical IPPSF is perfused at 60 to 80 %RH at 37°C. Viability studies described below were conducted in a recirculating mode. The medium is gassed with 95% oxygen and 5% carbon dioxide using a silastic-tube oxygenator. The perfusion media used for the IPPSF consists of a Krebs-Ringer bicarbonate buffer solution (pH 7.4) containing the chemicals listed in Table 1. A variable rate peristaltic pump circulates a total of 250 ml of media to the cannulated artery of the skin flap. To provide adequate oxygenation of the medium, a higher flow rate was used for the arterial-venous shunt line. Mixing is provided using magnetic stir bars in the arterial and venous reservoirs. In order to maintain arterial glucose concentrations between 80 to 120 mg/dl and a perfusate pH of 7.4, glucose and sodium bicarbonate are periodically infused into the shunt line. A microcomputer displays and stores arterial perfusate pressure and temperature. Perfusate flow and pH are continuously monitored and entered into the computer. At the termination of each IPPSF experiment, skin samples were collected for both light microscopy (LM) and transmission electron microscopy (TEM) to assess the morphological viability of the flap.[12]

B. CHARACTERIZATION OF EPIDERMAL VIABILITY
1. Morphological Studies
    If a morphological viability index was to be developed for the IPPSF, viability must first be defined. Viability in an IPPSF has a different connotation from that employed in cell culture work. In the flap, the majority of cells must be metabolically active, while in cell culture, only a small percentage of cells must be capable of subsequent growth on culture medium, and cells may be in various dormant phases. Therefore, the criteria for viability in the IPPSF will be more stringent than for other systems. Morphologic changes due to cell death, surgery, isolated perfusion, or chemically induced toxicity had to first be differentiated before a meaningful index could be established.
    Specimens for LM are fixed overnight in half-strength Karnovsky's fixative at 4°C (2% paraformaldehyde and 2.5% glutaraldehyde, approximately 976 mOSM in 0.1 *M* cacodylate

buffer), routinely processed, embedded in paraffin, and stained with hematoxylin and eosin (H&E). For TEM, tissue samples were minced into 1 mm³ and fixed overnight in half-strength Karnovsky's (4°C, pH 7.4, 976 mOSM), postfixed in 1% osmium tetroxide for 1 hour, dehydrated through graded ethanol solutions and infiltrated and embedded in Spurr's resin, placed in flat embedding molds and polymerized in a 70°C oven overnight.[1,13] Thick sections (approximately 1 μm) were stained with 1% toluidine blue in 1% sodium borate for orientation. Ultrathin sections approximately 600 Å were sectioned with a diamond knife. Sections were picked up on 300 mesh copper grids and stained with uranyl acetate and lead citrate and sections were then examined on a Philips 410LS transmission electron microscope (Philips, Mahwey, NJ).

a. Cell Death over Time in Detached Skin

In order to define changes due to cell death, a 4 cm × 12 cm piece of skin was harvested from the normal surgical site. Samples of skin were pinned to dental wax in a dissecting tray and floated on a 37°C water bath. The samples were kept moist only on the dermal side by bathing with nonoxygenated lactated Ringer's in which antimicrobials were added (10 μg/ml gentamicin and 0.2 μg/ml amphotericin B). Six millimeter biopsy samples were taken from different locations on the flap based on a random digit table at 0, 15, 30, and 45 min and 1, 2, 3, 4, 5, 6, 8, 9, 10, 12, 18, 24, 48, and 72 h in 5 pigs. The 6 mm biopsy was further divided in half for LM and other half for TEM. Quantitative measurements on dark basal nuclei, pyknotic basal cells, basal vacuoles, dark stratum spinosum nuclei, pyknotic stratum spinosum, and stratum spinosum nuclei, were first based on a scale from 1 = (0—5 cells), 2 = (5—10), 3 = (10—20), 4 = (20—40), 5 = (diffuse). The second assessment was a subjective rating on the overall appearance of a section, scored as 1 (no change—normal), 2 (mild changes), and 3 (obviously necrotic). This rating was found to correlate to the individual assessment.

LM evaluation on H&E sections collected from dying skin revealed noticeable changes occurring over time. A discriminant analysis revealed that the scores based on the counts of pyknotic basal cells, basal vacuoles, and pyknotic stratum spinosum contained information sufficient to correctly classify cell death. Hence the sum of these three scores was used as an appropriate morphologic viability index, a decrease in viability being reflected as an increase in the index. The shift from 1 to 2 (normal to transition) occurred as early as 3 h. Similarly, the shift from 2 to 3 (transition to necrotic) occurred as early as 12 h. Obvious abnormal LM changes occured at 12 h consisting of dark basal nuclei, pyknotic basal cells, basal vacuoles, dark stratum spinosum cells, pyknotic stratum spinosum cells, and stratum spinosum vacuoles. By 24 h, some pyknosis and degenerative changes occurred within the epidermal cell layers. At 48 h, the dermis was abnormal, being granular in appearance, and almost all the epidermal cells were pyknotic, and epidermal-dermal separation had occured. TEM observations showed normal morphologic integrity until 8 h. At 8 h, morphological alterations consisted of large single vacuoles, disruption of mitochondrial membranes, chromatin-clumping with nucleolar segregation, and nucleolar pleomorphism occurring within the stratum basale and stratum spinosum layers. At 12 h (Plate 4A*), nuclear envelope separation, single vacuoles, separation of desmosomes and degenerative organelles and cellular debris were in the intercellular space, yet the stratum corneum remained intact. By 24 h, classical pyknosis occurred with shrinkage of the nucleus and condensation of chromatin. Other changes associated with necrosis such as karyorrhexis, larger vacuoles, and at times lipid droplets near the mitochondrion were observed.[14]

b. After Surgery of Stage I Procedure (In Situ)

In order to define changes that occur in *in situ* flaps post-surgery, a biopsy study was performed *in situ* for 7 d after creating the flap in the stage I procedure. Morphometric

---

* Plate 4 follows p. 184.

TABLE 2
Morphometric Analyses of
Epidermal Thickness[a] in *In*
Situ **Skin Flaps (n = 3) At**
Various Times Between Stage
I And Stage II Surgeries

Day	Thin areas (μm)	Rete pegs (μm)
0	37.1 ± 2.8[b]	65.8 ± 16.9
1	43.5 ± 2.8	80.0 ± 13.0
2	68.1 ± 6.8	102.3 ± 4.4
3	82.9 ± 15.6	141.3 ± 12.0
4	67.4 ± 12.7	178.4 ± 13.8
5	96.4 ± 7.2	165.4 ± 11.8
6	86.2 ± 16.0	169.1 ± 23.5
7	68.1 ± 11.0	152.9 ± 38.3

[a] Stratum corneum thickness did not vary
and was not included in the epidermal
measurements
[b] Mean ± SE in three *in situ* flaps.

analyses of LM samples were performed on three *in situ* skin flaps to help determine the optimal time of harvest following stage I surgery. Biopsies were collected daily on the flaps, from 0 (30 min after stage I surgery) to 7 d, at random sites along the length of the skin flap whose free tip was sutured to the body wall to prevent traumatic injury. Pigs were immobilized with xylazine and ketamine to effect. During these sampling procedures, the pig felt no pain, since cutaneous innervation was severed during the formation of the tubed flap. Paraffin-embedded sections, stained with H&E, were examined with a $40 \times$ objective using a calibrated eyepiece reticle. Each section was evaluated at four points, two each at thin epidermal areas and two at the thickest areas, coinciding with rete pegs. Measured thickness of the stratum corneum was constant at all times. In the *in situ* study, LM at day 0 after surgery demonstrated that the epidermis and dermis were normal. At day 1, edema was present in the superficial and deep dermis and chromatin-clumping was seen in the epidermal cells. Day 2, edema was present in both superficial and deep papillary layers of the dermis and lymphatic vessels were dilated. Slight intracellular edema and chromatin clumping were present in the epidermal cell layers. Perivascular to diffuse eosinophilic mononuclear infiltrate occurred in the dermis. On day 3, similar changes were noted but dark stratum basale cells were seen, the significance of which is not known. Days 4, 5, and 6 were similar. By day 7, very slight intracellular edema of the epidermal cells was seen, perivascular and diffuse eosinophilic mononuclear infiltrates were present, but the superficial and deep dermis appeared normal.

From the morphometric analysis (Table 2), it is clear that the major difference attributable to the time between stage I and stage II surgeries is a thickened epidermal layer. By 7 d after stage I surgery, epidermal thickness had approximately doubled in both thin areas and at rete pegs, in addition to having greater variability over time, suggesting that the 2-d preparation may be more appropriate for percutaneous absorption studies. These data are also consistent with the hypertrophy seen in most 6-d IPPSF's. Ultrastructural observations at day 0 and day 1 were normal. On day 2, redistribution of nucleolar components into fibrillar and granular ribonucleoprotein filaments occurred. Several cells in mitosis were present in the stratum basale and spinosum layers. Some cells possessed a reticular nucleolonema with light granular and dense filamentous components and rounded pars amorpha areas or fibrous centers (Plate 4B). On day 3, the ultrastructure of the nucleolus resembled that on day 2 but tended to have an increase in the dense filamentous component. On day 4, irregularities developed within the nucleolus — that is, an irregular shape, a compact appearance and nucleolar margination. On day 5, the nucleolus was enlarged and consisted of dense filamentous components. On day 6, nucleolar margination was prominent and the nucleolus was enlarged. Several cells of the stratum basale and spinosum were characterized by a compact appearance without a developed nucleolonema and uniform distribution of structures containing RNA. Nucleoli of some other cells consisted of a dense fibrillar center with a periphery of granular components. On day 7, the nucleoli of both the stratum basale and stratum spinosum cells resembled that of day 6. No nucleolar caps, microsegregation or macrosegregation was noted.

c. Isolated Perfusion

In order to define changes specifically related to isolated perfusion, IPPSF tissue samples were taken after 12 h of perfusion.

LM of the epidermis in IPPSF's demonstrated normal intact epidermis (Plate 5*). To assess individual cellular morphology, TEM was also employed. Nucleolar pleomorphism was present in the stratum basale and spinosum layers (Plate 4C). Some vacuolization was also observed, depending on the location of the sample on the IPPSF. Also, some intercellular edema was evident. However, degenerative changes such as those associated with cell death in the control detached skin study were not evident.

d. Sodium Fluoride Toxicity

In order to determine whether the IPPSF was responsive to chemically induced toxicity, 10 mg/ml of sodium fluoride (NaF) was administered to four flaps after 5 and 12 h of stable perfusion. Toxicity was evidenced by the cessation of glucose utilization (see following section for method of analysis) 40 to 90 min later.

TEM on IPPSFs receiving NaF revealed normal epidermis except for the presence of large multiple vacuoles, often membrane-bound, and at times showed an amorphous substance in the stratum basale and spinosum layers (Plate 4D). These changes seen in the NaF flaps should be considered acute toxic lesions rather than degenerative changes secondary to loss of viability.[14]

2. Biochemical Studies

Hourly samples of both arterial (1.0 ml) and venous (1.5 ml) perfusate were collected to determine glucose concentration (mg/dl) (Glucose Analyzer 2; Beckman, Fullerton, Calif.) and osmolality (mOsm/kg) as well as venous lactate and lactate dehydrogenase (LD) concentrations. Glucose utilization (mg/h) was calculated from the product of glucose extraction (mg/dl), and the flow rate (ml/min) at each observation time. Cumulative glucose consumption was estimated from the area under the glucose clearance vs. time curve using the trapezoidal rule.[12] Lactate was assayed using a commercial method (Sigma Chemical Co., St. Louis, MO) and LD activity was determined by using a multistat centrifugal analyzer (MCA III+; Instrumentation Laboratories, Lexington, MA). Lactate was used in combination with glucose utilization to assess metabolic activity, and LD (an intracellular enzyme) leakage into the perfusate was used as a marker of cellular integrity.

One of the characteristics of a morphologically viable skin flap is a glucose utilization of 20 to 40 mg/h for a typical 30-g skin flap (0.7 to 1.3 mg/g-hr; 0.2 mg/cm²-h).[15] Cumulative glucose utilization is linear over the course of an experiment as depicted in Figure 3. Nonviable preparations have glucose utilizations less than 10 mg/h. Lactate production is linearly related to glucose utilization with a molar lactate to glucose ratio of 1.7. This finding agrees with numerous studies on cutaneous glucose utilization which suggest that 70 to 80%

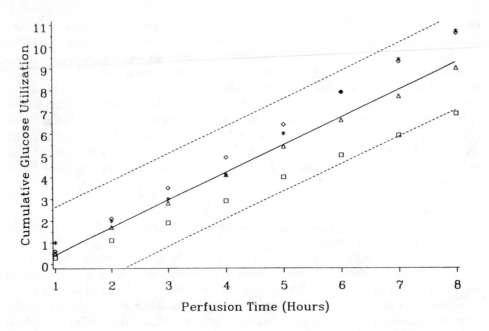

FIGURE 3.   Cumulative glucose utilization (mg/h-g of tissue) versus perfusion time based on four untreated/ normal skin flaps.

of cutaneous glucose is metabolized via glycolysis with lactate as the primary end product.[16,17] Terminal LD concentrations in the arterial media perfusing the viable preparations are less than 10 IU/l.

IV. APPLICATION OF THE IPPSF TO PERCUTANEOUS ABSORPTION

The IPPSF is a useful and unique model for studying percutaneous absorption of xenobiotics. This model allows for experimental manipulation of drugs, perfusate flow, pH, temperature, and ambient humidity. The independent and accessible vasculature in this model allows for quantitating the bioavailability of topically applied compounds or studying their cutaneous metabolism. If the compound being studied has vasoactive properties, the effect of vasodilation or vasoconstriction on percutaneous absorption can be directly measured. Also, the model can assess the distribution of systemically administered drugs to skin by modeling uptake of drug after arterial infusion into the skin flap.

To date, our laboratory has perfused over 700 IPPSFs for studying perfusion technique, assessing biochemical and morphological viability, assessing cutaneous toxicity of topically applied compounds, studying cutaneous uptake of infused drugs (viz. cisplatin, carboplatin, tetracycline, doxycycline), and modeling of the percutaneous absorption of 14 compounds including benzoic acid, caffeine, chlorobenzilate, cyclosporine, diisopropylfluorophosphate (DFP), lidocane, lindane, malathion, parathion, progesterone, testosterone, and triamcinolone.[18] Before details of these experiments are presented, the method for topically applying compounds will be discussed.

A. APPLICATION FOR DOSING

Once the skin flap has been in the perfused chamber for a period of 1 to 2 h (trial period), and the glucose utilization is greater than 10 mg/h, the flap is then considered to be viable based on this previously described biochemical indicator.

For topical dosing, the flap is removed from the apparatus and fitted with an 8 cm ×

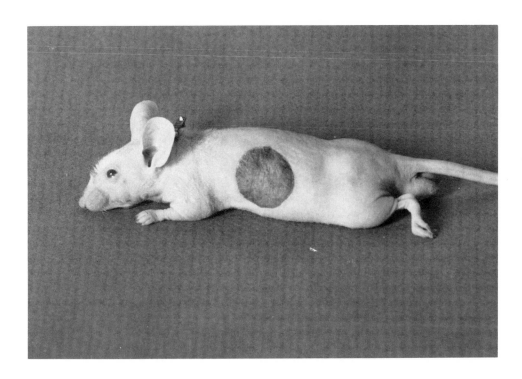

PLATE 1. Human split thickness skin graft on nude mouse

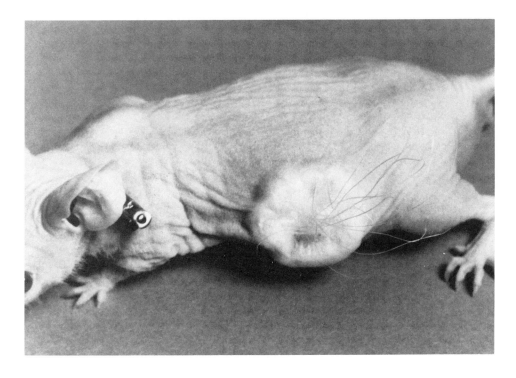

PLATE 2. Pig split thickness skin graft on nude mouse exhibiting pig bristle growth

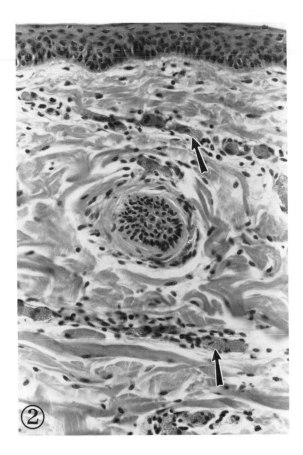

PLATE 3. Light micrograph from the center of a tubed skin flap. Note the normal epidermis, and dermis. Barium perfusate is evident in the superficial and deep capillaries of the dermis (arrows, H&E, 200×).

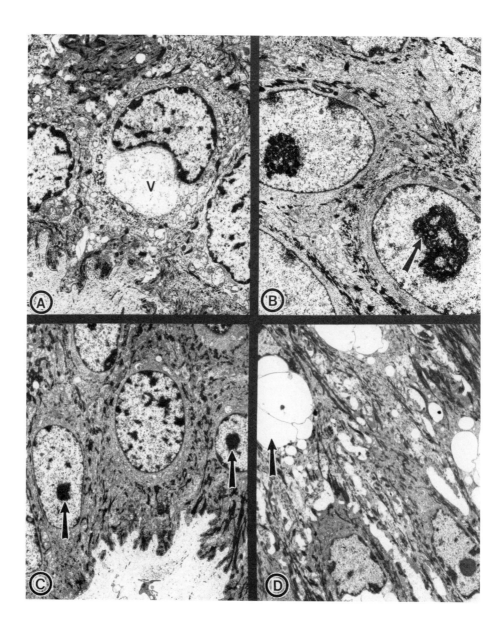

PLATE 4.   (A) Electron micrograph of detached skin after twelve hours. Note nuclear envelope separation and single vacuoles (v) 5500×. (B) Electron micrograph of a flap 2 d post-surgery (*in situ*). Note the nucleolar redistribution (arrow) 5720×. (C) Electron micrograph showing normal epidermal-dermal junction and epidermal cell layers after 12 h of perfusion. Note nucleolar pleomorphism (arrows) in the stratum basale cells. 3120× (Reprinted from Monteiro-Riviere et al.[14]) (D) Electron micrograph of a flap administered sodium fluoride. Note the multiple vacuoles (v) 3000×.

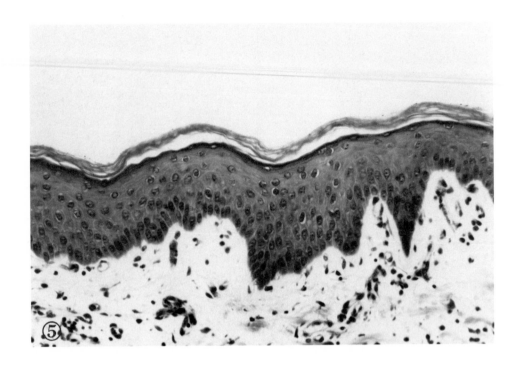

PLATE 5. Light micrograph of an IPPSF after 12 h of perfusion 330×.

3 cm patch of flexible plastic (Stomahesive; Convatec, Princeton, NJ) from which a 5 cm × 1 cm central area is cut out as the intended dosing area. This patch is glued to the skin flap with a non-toxic adhesive (Skin Bond; Pfizer Hospital Products, Largo, FL). The solutions are applied topically to this 5 cm^2 dosing area using a microsyringe to evenly distribute the designated solution over the exposed skin flap surface. Then the skin flap is returned to the apparatus after dosing.

For absorption studies requiring a nonocclusive patch, an additional aluminum foil bridge is glued over the patch and the flap is wrapped lightly in gauze and then returned to the perfusion apparatus. For studies requiring occlusion, the patch is wrapped in cellophane. If a transdermal drug delivery system is used, the patch is simply put on the IPPSF in the identical manner as used *in vivo*. The large surface area of the IPPSF is a major advantage for its use as an *in vitro* model.

B. PERCUTANEOUS ABSORPTION

In order to illustrate the application of the IPPSF to percutaneous absorption, Figure 4 depicts the observed venous flux profiles after topical application of ^{14}C benzoic acid (n = 6) and caffeine (n = 7) at a concentration of 40 $\mu g/cm^2$ (Mean ± SE).[19] One can readily appreciate the more rapid absorption of benzoic acid relative to caffeine. The increased variance seen at later time points with caffeine is due to the occurrence of increased perfusate flow seen in some IPPSFs, presumably due to a caffeine-induced vasodilation. In order to quantitate the extent of percutaneous absorption, these profiles were fitted to a three compartment linear pharmacokinetic model fully described elsewhere.[19] The IPPSF predicted bioavailability using the model derived parameters, of benzoic acid, caffeine, and 5 other compounds at 6 d was compared to the *in vivo* bioavailability determined using identical dosing conditions and is depicted in Figure 5. The correlation coefficient of $R^2 = 0.88$ (R = 0.94) shows the excellent predictive ability of the IPPSF. If compounds with slow absorptive flux are to be studied, drug may be applied to the skin flap *in situ* and IPPSFs harvested at any time post drug application. This approach removes any time limitations on experimental design.

The simple three compartment model utilized in these studies resulted in an *in vitro* to *in vivo* correlation which had a significant positive slope. Utilization of a hybrid physiological-compartmental pharmacokinetic model to fit this data is theoretically more sound since the cutaneous microcirculation can be modelled directly.[18,20,21] Using this approach, the *in vitro* to *in vivo* correlation was similar and the slope of the regression was close to one.

Another application demonstrating the utility of the IPPSF for percutaneous absorption involved studies with lidocaine iontophoresis.[18,22] Iontophoresis is a drug delivery system whereby electric current is used to drive ionic compounds across the skin. The IPPSF was used to determine whether coiontophoresis of a vasoactive compound with lidocaine would modulate the venous flux of drug. As seen in Figure 6, when the vasoconstrictor epinephrine was simultaneously delivered with lidocaine, the peak flux of lidocaine was reduced and the retention time in the skin was prolonged. In contrast, coiontophoresis of the vasodilator tolazoline resulted in an increased transdermal flux of drug. These experiments illustrate one of the major advantages of the IPPSF as an *in vitro* model for assessing percutaneous absorption since compounds which alter the microcirculation can be directly assessed. Other *in vitro* systems which do not have an intact vasculature cannot study this clinically relevant phenomenon.

C. CUTANEOUS BIOTRANSFORMATION

The IPPSF has also been utilized to study the cutaneous biotransformation of topically applied parathion[23] and chlorobenzilate. These experiments are conducted by applying radiolabeled compound as described above and then serially monitoring venous effluent for

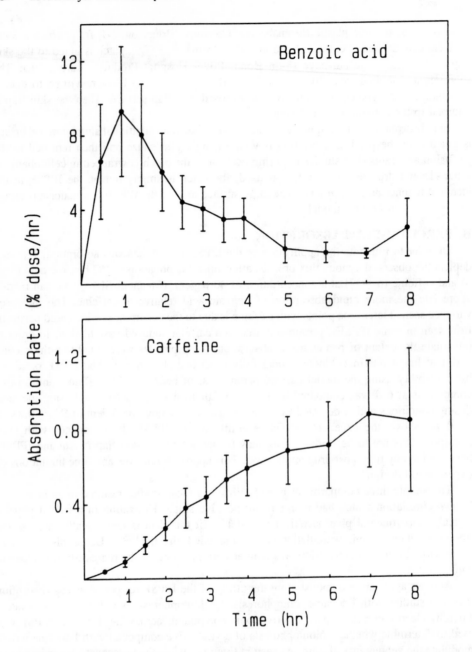

FIGURE 4.   Absorption rate-time curves of radioactivity for benzoic acid (B) and caffeine (C). Each point is the mean of 6 (B) or 7 (C) separate experiments and the bars represent ± 1 SE. (Reprinted from Carver et al.[19]).

the presence of parent compound and metabolites. After parathion application, approximately 70% of the absorbed radiolabel recovered was paraoxon and 5% was paranitrophenol. When the flap was pretreated by adding the nonspecific cytochrome monooxygenase inhibitor 1-aminobenzotriazole to the perfusion media, paraoxon formation was effectively blocked. Similarly, occlusion of the application site also significantly affected the observed metabolic profile. When the organochlorine pesticide chlorobenzilate was applied, a significant fraction of the absorbed dose was present as either the phenone or as one unidentified metabolite.

These studies demonstrate another major factor for utilizing a viable perfused skin preparation for absorption studies since detection of first-pass cutaneous biotransformation

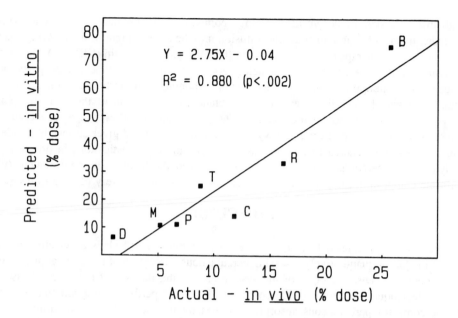

FIGURE 5. Correlation of IPPSF *in vitro* to *in vivo* absorption using seven compounds applied at 40 μg/cm² nonoccluded in an ethanol vehicle in both systems: caffeine (C), benzoic acid (B), malathion (M), parathion (P), testosterone (T), progesterone (R) and DFP (D). (Reprinted from Carver et al.[19])

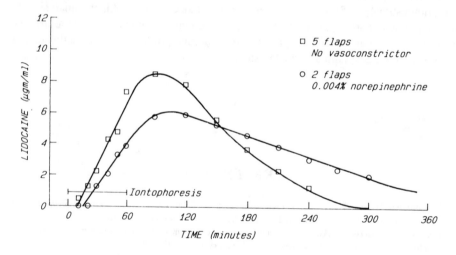

FIGURE 6. IPPSF venous flux profile of 1% lidocaine administered by iontophoresis (5 cm², 2 mA). (Reprinted from Riviere et al.[18]).

is facilitated by simply monitoring cutaneous venous flux of compound. Future experiments will study the effects of altered environmental conditions and metabolic state on cutaneous metabolism of xenobiotics.

D. CUTANEOUS UPTAKE

The final application of the IPPSF to cutaneous pharmacology involves modelling the cutaneous uptake of drug infused into the arterial perfusate. To study this process, a mathematically identifiable hybrid physiologic compartmental pharmacokinetic model is used to predict uptake of drug into skin after infusion into the arterial input of the IPPSF.[20] Using the cancer chemotherapeutic agents cisplatin (CDDP) and carboplatin (CBDCA), the extent of tissue extraction over time was modeled. As predicted from their chemical properties, tissue accumulation of cisplatin exceeded that of carboplatin. Similarly, when the antibiotics tetracycline and doxycycline were studied, cutaneous distribution of the more lipophilic doxycycline exceeded that of tetracycline. These experiments allow one to quantitate the rate of cutaneous uptake of drug after systemic administration and gives an insight into how drugs can be administered to optimize skin concentrations. Investigations in progress are studying how alterations in cutaneous blood flow and the local tissue environment (pH, perfusate composition, hyperthermia, disease) affect systemic delivery of drug to the skin.

V. DISCUSSION

As can be appreciated from the above presentation, the IPPSF is a versatile *in vitro* animal model for quantitatively assessing percutaneous absorption of topically applied drugs and xenobiotics. Disadvantages of the system are (1) the flap is not reusable, (2) aseptic surgical techniques are required, (3) cost, and (4) limited perfusion time although this can be overcome for percutaneous absorption studies by *in situ* application of drug prior to harvest. The major advantages of the IPPSF include: (1) an isolated system with control over physiological parameters and perfusate composition, (2) an intact functional microcirculation responsive to topically applied compounds, (3) a viable epidermis allowing cutaneous biotransformation to be assessed, (4) a predictable extrapolation to *in vivo* percutaneous absorption estimates, (5) a system amenable to detailed pharmacokinetic analysis, (6) skin (porcine) which is structurally and functionally similar to human skin, (7) a large surface area of skin which allows human prototype transdermal products to be directly studied, and finally (8) an humane animal model system.

REFERENCES

1. **Monteiro-Riviere, N. A.,** Ultrastructural evaluation of the porcine integument, in *Swine in Biomedical Research,* Vol. 1, Tumbleson, M. E., Ed., Plenum Press, New York, 1986, 641.
2. **Monteiro-Riviere, N. A.,** Ultrastructure of the integument of the domestic pig *(Sus scrofa)* from one through fourteen weeks of age, *Anat. Histol. Embryol.,* 14, 97, 1985.
3. **Monteiro-Riviere, N. A.,** Comparative anatomy, physiology and biochemistry of mammalian skin, in *Fundamentals and Methods of Dermal and Ocular Toxicology,* Hobson, D. W., Ed., The Telford Press, New Jersey, 1990, chap. 1.
4. **Forbes, D. P.,** Vascular supply of the skin and hair in swine, in *Advances in Biology of the Skin,* Montagna, W. and Dobson, R., Eds., Pergamon Press, New York, 1969, 419.
5. **Ingram, D. L. and Weaver, M. E.,** A qualitative study of the blood vessels of the pig's skin and the influence of environmental temperature, *Anat. Rec.,* 163, 517, 1969.
6. **Meyer, W., Neurand, K., and Radke, B.,** Elastic fibre arrangement in the skin of pigs, *Arch. Dermatol. Res.,* 270, 391, 1981.
7. **Meyer, W., Neurand, K., and Radke, B.,** Collagen fibre arrangements in the skin of pigs, *J. Anat.,* 134, 139, 1982.
8. **Meyer, W., Schwarz, R., and Neurand, K.,** The skin of domestic mammals as a model for the human skin with special reference to the domestic pig, *Curr. Probl. Dermatol.,* 7, 39, 1978.
9. **Wester, R. C. and Maibach, H. I.,** Cutaneous pharmacokinetics: 10 steps to percutaneous absorption, *Drug Metab. Rev.,* 14, 169, 1983.

10. **Wester, R. C. and Maibach, H. I.,** Advances in percutaneous absorption, in *Cutaneous Toxicity,* Drill, V. A. and Lazar, P., Eds., Raven Press, New York, 1984, 29.
11. **Bowman, K. F., Monteiro-Riviere, N. A., and Riviere, J. E.,** Development of surgical techniques for preparation of *in vitro* isolated porcine skin flaps for percutaneous absorption studies, *Vet. Res.,* (In Press).
12. **Riviere, J. E., Bowman, K. F., Monteiro-Riviere, N. A., Dix, L. P., and Carver, M. P.,** The isolated perfused porcine skin flap (IPPSF). I. A novel *in vitro* model for percutaneous absorption and cutaneous toxicology studies, *Fund. Appl. Toxicol.,* 7, 444, 1986.
13. **Monteiro-Riviere, N. A. and Manning, T. O.,** The effects of different fixatives on the porcine integument, in *Proc. of the 45th Annual Meeting of the Electron Microscopy Society of America,* Bailey, G. W., Ed., San Francisco Press, San Francisco, CA, 1987, 948.
14. **Monteiro-Riviere, N. A., Bowman, K. F., Scheidt, V. J., and Riviere, J. E.,** The isolated perfused porcine skin flap (IPPSF). II. Ultrastructural and histological characterization of epidermal viability, *In Vitro Toxicol.,* 1, 241, 1987.
15. **Riviere, J. E., Bowman, K. F., and Monteiro-Riviere, N. A.,** On the definition of viability in isolated perfused skin preparations, *Br. J. Dermatol.,* 116, 739, 1987.
16. **Johnson, T. A. and Fusaro, R.,** The role of the skin in carbohydrate metabolism, *Adv. Metab. Disord.,* 6, 1, 1972.
17. **Frienkel, R. K.,** Carbohydrate metabolism of the epidermis, in *Biochemistry and Physiology of the Skin,* Vol. I, Goldsmith, L. A., Ed., Oxford University Press, New York, 1983, 328.
18. **Riviere, J. E. and Carver, M. P.,** Isolated perfused skin flap and skin grafting techniques, in *Fundamentals and Methods of Dermal and Ocular Toxicology,* Hobson, D. W., Ed., The Telford Press, New Jersey, 1990, chap. 10.
19. **Carver, M. P., Williams, P. L., and Riviere, J. E.,** The isolated perfused porcine skin flap. III. Percutaneous absorption pharmacokinetics of organophosphates, steroids, benzoic acid, and caffeine, *Toxicol. Appl. Pharmacol.,* 97, 324, 1989.
20. **Williams, P. L. and Riviere, J. E.,** Definition of a physiologic pharmacokinetic model of cutaneous drug distribution using the isolated perfused porcine skin flap (IPPSF), *J. Pharm. Sci.,* 78, 550, 1989.
21. **Williams, P. L. and Riviere, J. E.,** A pharmacokinetic model describing xenobiotic percutaneous absorption in the isolated perfused porcine skin flap, presented at the Second Annual Symposium on Frontiers of Pharmacokinetics and Pharmacodynamics, Little Rock, Arkansas, October 12 to 14, 1988, 34.
22. **Riviere, J. E., Sage, B. H., and Monteiro-Riviere, N. A.,** Transdermal lidocaine iontophoresis in isolated perfused porcine skin, *J. Toxicol.: Cut. Ocul.,* 8, in press, 1990.
23. **Carver, M. P., Levi, P. E., and Riviere, J. E.,** Significant first pass bioactivation of parathion (P) during percutaneous absorption in the isolated perfused porcine skin flap (IPPSF), *Toxicologist,* 8, 125, 1988.
24. **Riviere, J. E.,** personal communication, 1988.

Chapter 12

VALIDITY OF SKIN ABSORPTION AND METABOLISM STUDIES

John Kao

TABLE OF CONTENTS

I. INTRODUCTION

As the largest and most external organ, the skin is constantly exposed to the hazards of the environment, and it is often viewed as a living protective envelope surrounding the body. It serves as a barrier limiting the systemic exposure to the effects of harmful external agents, and prevents the excessive loss of critical internal constituents. However, it is becoming increasingly apparent that the skin is not a complete barrier; in fact it is a readily accessible portal, with a large surface area, through which a variety of chemical agents can enter the body, and subsequently pass into the systemic circulation.

It has long been recognized that skin exposure to chemicals can result in adverse health effects, and some 50% of all reported industrial illnesses are estimated to be related to dermal exposure to hazardous chemicals.[1] Indeed, the National Institute for Occupational Safety and Health has stated that "skin disorders resulting from exposure to industrial chemicals are the most pervasive current occupational health problem in the U.S."[2] This increased awareness of the toxicological consequences of dermal exposure has highlighted the inadequacies of our current knowledge concerning skin function as it relates to skin absorption and cutaneous toxicity of chemicals. Consequently, skin absorption studies have been an important facet of toxicological investigations. On the other hand, our knowledge of skin permeability and understanding of some important fundamentals governing skin absorption have resulted in a dramatic surge of interest in the transdermal route as a means of controlled delivery of potent drugs for systemic, as opposed to local therapy. The success with transdermally delivered nitroglycerin has provided much of the impetus for the development of this technology, and recent advances include transdermal systems for drugs such as scopolamine, clonidine, and estradiol.

Until recently, research activities in skin absorption were concerned primarily with physiochemical and biophysical factors that influence skin penetration and permeation.[3-16] However, advances in skin biology have resulted in the discovery of novel, but important physiological functions and metabolic capabilities in this organ, and some of these biochemical properties may have important functions in skin absorption. Studies with various skin preparations have shown this organ to be capable of performing a variety of metabolic functions including those involved in the metabolism of hormones, drugs, and environmental chemicals. Indeed the skin, with its complement of drug metabolizing activities, is now recognized as an important organ for the extrahepatic metabolism of xenobiotics.[17-21] Since skin contains the enzymes responsible for metabolizing xenobiotics, any chemicals that are applied to the skin surface will, during the course of penetration and permeation through this organ, be exposed to any available biotransformation systems that are present in the skin. Consequently, inactivation, activation and interaction with tissue components may occur during the translocation of the chemicals across the skin. Therefore, questions concerning the functional significance of skin metabolism in the percutaneous fate of chemicals are of considerable interest in both dermatotoxicology and dermatopharmaceutics: To what extent are topically applied chemicals metabolized by the skin during percutaneous absorption? Can cutaneous metabolism influence skin absorption, and be a mediator in the development of either local or systemic toxicity? What are the factors that may modulate the metabolic capability of the skin, and indeed how may such modulation affect the physiological disposition of topically applied chemicals? These are just a few examples of scientific issues relating to percutaneous absorption that need to be addressed. From a pharmaceutical development point of view, will skin metabolism prove to be an opportunity, or a nuisance in the future development of novel transdermal drug delivery systems? To what extent will cutaneous metabolism alter the pharmacokinetics and bioavailability of topically delivered drugs? Will the magnitude of the metabolic process in the skin be sufficient to make the prodrug approach to controlled transcutaneous therapeutics a reality? It is clear that transdermal delivery of drugs offers many theoretical advantages over conventional therapy; but, a better understanding of the relevance of the poorly defined biochemical and toxicological factors that influence skin absorption and fate of topically applied agents would facilitate continued advances in this area.

Skin absorption is a complex phenomenon, and the fundamental concepts relating to this process have been extensively reviewed, and noteworthy reviews in the recent literature include the in-depth treatises of Schaefer et al.[22] and Barry.[23] Percutaneous absorption can be viewed as the translocation of surface applied substances through the various layers of the epidermis and dermis to a location where they can enter the systemic circulation via the cutaneous vasculature and lymphatics, or remain in the deeper layers of the skin. This transport of substances through the skin involves complex diffusional and metabolic processes, and is influenced by the interaction of a variety of physiochemical, biophysical, and biochemical factors. Absorption is the net result of the penetration, cutaneous metabolism, binding and permeation of a topically applied chemical into and through the different strata of the skin. Assessment of this process following topical application of drugs and environmental chemicals is becoming an increasingly important aspect of both toxicological and pharmaceutical investigations. Relative to toxicology, the ultimate aims of skin absorption studies are to identify and quantify the potential cutaneous toxicity, estimate the relative risk, and hopefully develop the appropriate strategies to minimize this risk resulting from topical exposure. In contrast, percutaneous absorption studies in transdermal delivery are designed primarily to assess and manipulate the rates of transport of drugs across the skin, and ultimately to determine if such rates are sufficient to achieve the desired exposure and provide optimal therapeutic response. The aim is to identify or design the therapeutic compound with the appropriate properties for commercial development. Closely associated with these studies are investigations that are designed to assess the cutaneous and systemic bioavailability and bioequivalence of the compounds under development. Thus, depending on ones perspective, the focus of skin absorption studies may be to increase penetration or to reduce absorption.

As described in detail in previous chapters of this book, a wide variety of experimental procedures have been developed and used to assess skin absorption. However, at present, there are no generally accepted techniques that are completely satisfactory for assessing skin absorption. Debates and conflicting opinions continue to revolve around the various factors that are important in influencing percutaneous absorption; consequently, the rationales by which experimental designs are selected and developed are continually being modified and redefined. The fundamental questions addressed in most skin absorption investigations are concerned with how much, how fast and what are the modulating factors that may influence the penetration and percutaneous fate of topically applied agents. The methods available to answer these questions are broadly divided into *in vitro* and *in vivo* studies. Both approaches have their advantages, limitations and weaknesses. The purpose of this chapter is to compare and contrast the various methods employed for dermal absorption by examining the principles and practices that govern the different experimental approaches, and to assess their relative limitations from the perspective of extrapolation to the real world situation.

II. LIMITATIONS IN SKIN ABSORPTION AND METABOLISM STUDIES

A. EXPERIMENTAL APPROACHES AND THEIR LIMITATIONS
1. *In Vitro* Studies

Diffusion laws form the theoretical basis of *in vitro* studies, and in general these studies are conducted with the excised skin preparation as the diffusion membrane separating the donor and receptor compartments of a diffusion chamber. Typically, an appropriate fluid is placed into the receptor compartment, and the compound under investigation, usually radiolabeled, in an appropriate formulation is placed in the donor compartment; the recovery of radioactivity, or the chemical of interest, with time in the receptor fluid then provides an estimate of skin absorption. The justification of this methodology centers upon the generally accepted assumption that the stratum corneum is the principal rate limiting barrier in skin absorption.[5-8] Since the outermost layer of the skin is composed essentially of nonliving tissues, it is reasoned therefore that biochemical processes are unlikely to influence the diffusional characteristics of the rate limiting membrane, and hence, *in vitro* diffusion studies will accurately measure skin penetration and absorption. The outcome of this assumption is that problems of percutaneous absorption are often simplified to a number of diffusion equations, and by carefully defining what are believed to be the fundamental physiochemical parameters underlying the diffusion process, the relative rates and extent of *in vivo* skin absorption of chemicals were often predicted and correlated on the basis of their mass transfer coefficients determined from *in vitro* studies. However, by considering the skin as merely a diffusional membrane, concepts relating to biochemical factors which may influence the percutaneous fate of topically applied substances have in the past received little attention. Fortunately this neglect is being addressed, and although investigations in this area are still in their infancy, recent studies have shown that skin metabolism can influence the percutaneous fate of certain chemicals.[24-28]

Experimental designs of *in vitro* studies can be distinguished in two strategies. In the traditional "steady state" or "infinite dose" technique,[22,23] a well-stirred donor solution of the compound of interest, at a defined constant concentration, is used to deliver the compound across the skin preparation, and the compound is received by a well stirred receptor. The most important design features of these studies are that the quantity of the compound that penetrates the membrane must be kept small relative to the total amount available; i.e., there is no appreciable reduction in the concentration of the compound in the donor compartment. This is so that steady-state or zero-order flux conditions are not significantly violated, and the studies are performed with rigorous compliance to the laws of diffusion. Under these experimental conditions, this methodology can provide information on flux, lag times and relevant mass transfer coefficients that are pertinent to steady state diffusion, and at steady-state flux conditions, the diffusional characteristics of a given compound, in a given vehicle through a defined membrane preparation can be characterized in terms of its permeability coefficients.[11,12,23] This approach has contributed some of the most definitive insights into the physiochemical parameters that govern membrane diffusion, and has provided much of our current knowledge concerning the diffusional aspects of percutaneous penetration.

While this infinite dose technique has been invaluable in the development of transdermal drug delivery concepts, this methodology may be of limited value as a predictive model for assessing skin absorption *in vivo*. To mimic *in vivo* conditions, the so-called "finite dose" technique was developed.[29] This is essentially a modification of the traditional steady-state method. The important difference is that the skin preparation is supported over the receptor so that the epidermal surfaces are not covered, and the compound of interest is applied to the surface of the skin in a manner similar to exposure *in vivo*. Recent refinements, such as flow through designs for the receptor fluid, provide additional features that mimic the *in vivo* conditions.[25,30-32] This approach has received a great deal of recent attention since it is potentially a very powerful tool for studying factors that influence percutaneous absorption and may provide a means for assessing the skin absorption of toxic agents.

Regardless of the *in vitro* study design chosen, the fundamental assumption for *in vitro* absorption investigations is that the excised skin retains its barrier functions,[33] and it has long been recognized that the diffusional integrity of this preparation is of major importance in the outcome of *in vitro* studies. Since the stratum corneum is generally believed to be the rate limiting barrier, the integrity of this skin layer is of paramount importance. In diffusion terms relative to percutaneous absorption, flux is dependent upon the concentration gradient across the barrier membrane, and also the partition of the penetrant between the vehicle and the stratum corneum;[11,12,23,34] consequently solubility of the agent of interest in the vehicle used could influence the outcome of the permeation determination. The vehicle as an important factor influencing skin absorption has been discussed in detail by others;[22,23,35] but it is worth noting that solvents and surfactants[36] are often employed as vehicles in skin absorption studies, and some of these vehicles may increase skin permeation by changing or destroying the barrier properties of the stratum corneum.

Relative to the stratum corneum, it is often assumed that diffusion through the epidermis and the papillary dermis was much higher and would not significantly influence the overall estimate of *in vitro* skin penetration. While this may be a reasonable assumption for freely diffusible penetrants such as aqueous alcohol solutions,[9,10] it is becoming clear that, because of the thicknesses of the dermis and the epidermis relative to the stratum corneum, the much thicker dermis and the epidermis may serve as the receptor under *in vitro* conditions. Indeed, using a "horizontal slicing" technique it has been demonstrated that, for compounds such as hydrocortisone, significant accumulation in the dermal compartment was evident following topical application *in vitro*, but not *in vivo*.[37] By virtue of its relative volume, the dermis may retain permeating substances *in vitro* so that compounds which penetrate the stratum corneum readily, may appear to permeate the skin relatively slowly because of their long transit time in the dermis; consequently *in vitro* methodology may underestimate the overall skin absorption of a given compound.[22] On the other hand, the stratum corneum may not be the only source of diffusional resistance in the skin. For certain lipophilic chemicals, it has been suggested that the lipid rich stratum corneum may not be a barrier, but rather, it may act as a reservoir, and function more as a sponge, capable of retaining a quantity of lipophilic material limited only by the solubility of the substance in the sebaceous and intrinsic epidermal lipids.[24,25] The largely aqueous epidermal and dermal tissue may be the more relevant diffusional barrier for such lipophilic chemicals, and therefore, the thickness of the skin preparation used in an *in vitro* investigation may have an important influence on the outcome of *in vitro* percutaneous absorption studies.

The skin absorption process is generally assumed to terminate at the site of vascular entry, and the dermal microvasculature which is situated within the region immediately beneath the epidermis is routinely viewed as an efficient "sink" removing the penetrant.[22,23] Consequently, thin sections of "split-thickness" skin are often advocated as being the more appropriate membrane preparation for *in vitro* investigations since they may be more representative of the in vivo situation,[22,23,38-41] In using such epidermal sections for *in vitro* studies, it is critical that the integrity of the diffusional barrier be established. This is particularly important in preparations from haired skins because holes in the preparation corresponding to hair follicles are often one of the major problems encountered; especially when very thin sections of skin are used. Skin sections prepared so as to include the hair roots may be one approach to reduce this problem; but the increase in thickness of the membrane may compromise the subsequent permeation measurements. This is a dilemma, and individual investigators will have to rationalize the appropriate choice of membranes used based on the nature of the *in vitro* study, and the compound being investigated. However, it should be noted that evidence is accumulating which indicates that the cutaneous microvasculature does not always function as a perfect "sink", and a fraction of the permeating material can be delivered to the subcutaneous fat and underlying connective and muscular tissue.[42] Moreover, for certain compounds the skin appendages, such as hair follicles, may play a significant role in the overall skin absorption process.[43-53] Appreciation of some of these issues would greatly assist in choosing the appropriate skin preparation for *in vitro* investigations and hopefully provide a more critical assessment of the results.

Ideally, freshly excised and viable skin preparations should be used for *in vitro* investigations. Unfortunately, this is not always possible, particularly when skin preparations are of human origin. It is a common practice to use excised skin preparations which have been stored under reduced temperature and/or in a dessicated state for *in vitro* percutaneous absorption investigations. The assumption is that the barrier properties of the skin are unaffected by storage. However, recent studies have demonstrated that storage conditions can have a profound influence on the results of *in vitro* investigations, and the effects may vary from compound to compound.[54-59] Consequently, when circumstances dictate that storage of the excised tissue is necessary, an assessment of the effects of storage on the permeation of the specific compound under investigation must be determined so that the validity of the *in vitro* observations can be assessed.[60]

Although physical diffusion is the principal determinant modulating skin absorption, recent observations have strongly indicated that cutaneous metabolism and the metabolic status of the skin can have a major influence on the percutaneous fate of topically applied chemicals.[24-28,61-67] Indeed the viable epidermis as a metabolic barrier has been the subject of a number of derivations exploring the mathematical modeling of percutaneous absorption.[68-73] Therefore, in designing the *in vitro* skin absorption studies, the relevance of biochemical viability of the excised tissue will need to be considered. For example, in studies where skin metabolism of the topically applied compound is of special interest, the biological integrity as well as the structural integrity of the excised tissue will be important factors that may affect the validity of the resulting observations. Consequently, it will be necessary to establish that the *in vitro* conditions of the study are capable of maintaining the metabolic viability of the excised tissue during the course of the investigation. The desirable *in vitro* conditions employed can be established in separate experiments prior to the permeation investigation, and the various approaches used to demonstrate and assess the viability and structural integrity of the excised tissues preparations have been described previously.[74,75] It is perhaps worth noting that techniques for the cryopreservation of viable mammalian skin have been described in the literature.[76,77] Although the criteria used in these reports for demonstrating biological viability varies considerably, systematic assessments of these methods as potential means for "banking" skin preparations may provide useful opportunities for *in vitro* dermal studies.

As indicated earlier, local effects of the vehicle may affect the barrier properties of the skin; indeed the compound under investigation may produce local toxicity following topical application. Where exposure results in obvious damage to the stratum corneum, enhanced permeability is to be expected.[16] However, biochemical effects in response to biologically active chemicals may also influence the functional capabilities of the excised skin. Unfortunately very little is known regarding the potential influence of the local toxicity on cutaneous metabolism and skin absorption. Therefore, an assessment of local toxicity following topical application may be necessary to assist in the evaluation of the *in vitro* skin absorption. Various histological and biochemical approaches which can be used to assess local toxicity, both *in vivo* and *in vitro* have been described.[74,75,78] However, to avoid unnecessary complications it is suggested that whenever possible a no effect dose of the compound should be determined and used to study its cutaneous absorption and metabolism *in vitro*. In any event, knowledge of the physical and biological integrity of the skin preparation before, during and after the experiment would greatly facilitate the interpretation of the *in vitro* observations.

Traditionally, the basis for determining skin absorption *in vitro* is that the recovery of material of interest in the receptor compartment provides an accurate measurement of skin penetration and permeation. This measurement is dependent upon the ability of the compound of interest to partition from the skin preparation to the receptor fluid; consequently, the validity of the *in vitro* measurements will depend largely on the appropriateness of the receptor fluid used. Therefore, caution must be exercised in the interpretation of the *in vitro* observations. In designing *in vitro* studies, the partitioning characteristics of the compound of interest and that of the receptor fluid relative to the skin preparation can be of paramount importance and should be given due consideration. Unfortunately, the selection of receptor fluid has been largely empirical, and in the pioneering studies, distilled water was used as the receptor fluid.[9,10] Until recently, isotonic saline, which may or may not be buffered at pH 7.4 has been the overwhelming choice in the receptor fluid.[22,23] On the other hand, when cutaneous metabolism as well as the absorption of topically applied compounds were investigated, an appropriate culture medium was used to maintain the viability of the tissue, and in these studies the culture medium also served as the receptor fluid.[24-28] However, in studies with lipophilic chemicals which have limited water solubility, there are increasing concerns regarding the suitability of aqueous salt solutions as receptor fluid.[79,80] For hydrophobic chemicals the rate limiting step in percutaneous absorption *in vitro* may not be diffusion through the stratum corneum but rather partitioning from the skin to the receptor fluid. Consequently, the relevance of the *in vitro* determinations may be seriously compromised. Under *in vivo* conditions, these compounds may readily penetrate the stratum corneum and diffuse through the skin and, because of the solubilizing and emulsifying abilities of biological fluid, be taken away by the blood in the dermal vasculature. Therefore, skin permeability of these compounds, based on *in vitro* determinations, may be grossly underestimated. This inability of hydrophobic chemicals to partition into aqueous receptor fluids is a potential liability which may limit the value of this *in vitro* methodology for assessing skin absorption and identifying the appropriate receptor fluid will require serious consideration.

Various approaches have been utilized to manipulate the lipophilicity of the receptor fluid, and these have included the use of physiologically based fluids, as in supplementing the aqueous buffered salt solutions with proteins.[81,82] Nonphysiological means, such as the incorporation of solvents and surfactants into the aqueous receptor fluids, have also been employed,[83-85] and these methods have resulted in limited success in improving *in vitro* permeation; however, whether these improvements will result in *in vitro* observations that accurately reflect *in vivo* absorption remains to be established. It is evident that the ideal universal receptor fluid has yet to be identified. Viability of the excised skin may be important in the percutaneous fate of topically applied agents. When identifying an appropriate receptor fluid, it is clear that the medium chosen should be compatible with the maintenance of not only the structural integrity of the excised skin but also its biological viability. The ideal receptor fluid should, therefore, possess the appropriate physiochemical and biochemical properties that are desirable for *in vitro* percutaneous absorption and metabolism investigations. What these desirable properties may be remains to be established; it is suggested that physiologically based media would be more appropriate. Until the ideal receptor fluid has been identified, the selection of a receptor fluid for a particular compound remains empirical, and the relevance of the *in vitro* results will probably reflect the views of individual investigators as to the relative role of the structural and biochemical functions of the skin as an interface in percutaneous absorption.

Since the basis of *in vitro* absorption measurements is generally the recovery of the compound under investigation in the receptor fluid, much of the research activity has naturally focused in this area, and recovery of material in the skin itself has received only limited attention. Cutaneous distribution, metabolism, and binding of the topically applied agent are integral parts of the percutaneous absorption process,[22,23] so that assessing the disposition of the applied chemical in the skin tissue should be an important measurement in the evaluations of skin absorption. Indeed, the amount of a chemical that passes through the stratum corneum into the viable epidermis and dermis is an important parameter for assessing local bioavailability, and it also contributes to the overall estimate of *in vitro* percutaneous absorption. Furthermore, analysis of the skin following permeation studies would assist in determining mass balance and dose accountability; however, such measurements are often not reported or conducted even though they are important in establishing the validity, and hence, the interpretation of *in vitro* observations. In the horizontal slicing technique,[37] in which the material of interest in the different layers of the skin may be determined, an experimental design is made available that permits the quantifications of the penetrant in the skin. Moreover, the methodology brings an additional dimension to *in vitro* absorption studies that should contribute to a better understanding of the percutaneous absorption process.

Because of obvious advantages, radiolabeled chemicals are routinely used in skin absorption studies, and frequently liquid scintillation spectrometry is the sole method used for detecting the pentrating substances in the receptor fluid. However, skin absorption may be accompanied by cutaneous metabolism, therefore the radioactivity recovered in the receptor medium reflects not only the permeation of the parent substance but also its metabolites. In view of our lack of understanding of the functional significance of cutaneous first pass metabolism in percutaneous absorption, identifying the permeating chemical entities with other analytical methods is increasingly becoming an integral part of *in vitro* skin absorption studies.

However, in assessing the significance of metabolism in skin absorption based on *in vitro* observations, caution must be exercised. As alluded to earlier regarding the *in vitro* permeation of lipophilic agents, it has been suggested that under the routinely employed *in vitro* conditions a lipophilic penetrant may be unable to freely partition from the skin to the aqueous receptor. Consequently, the compound may accumulate, and the longer residence time in the skin tissue may provide the viable epidermis and dermis a greater opportunity to convert the water-insoluble compound to water-soluble metabolites that are capable of partitioning freely into the receptor. Therefore, the recovery of significant amounts of metabolites in the receptor may be artifactual, and the influence of skin metabolism overestimated. Also, metabolites found in the receptor are often assumed to be the product of cutaneous metabolism, formed during permeation through the skin. However, it has been suggested that these metabolites may result from transformation in the receptor fluid after skin permeation. The inherent instability of the parent compound in the receptor could be a possible reason for these chemical transformations, for example, hydrolysis. Alternatively, these transformations may be mediated by active enzymes that have leaked into the receptor from the excised tissue, and metabolites are formed in the receptor fluid as artifacts of the *in vitro* methodology. This could be a difficult problem where soluble enzymes such as skin esterases are involved.[86] Therefore, in assessing the relevance of cutaneous metabolism in skin absorption *in vitro*, it will be important to determine the origin of the metabolites found in the receptor since enzyme mediated transformation after permeation could contribute significantly to the overall recovery of metabolites in the receptor fluid.

From the above discussion, it is evident that the *in vitro* methodology of skin absorption, which on the surface appears to be relatively simple, is shrouded with various limitations. While the principles of *in vitro* percutaneous absorption are well defined, in practice, the design and conduct of *in vitro* studies are complex, and interpretation of the results obtained require careful consideration. Nevertheless, appropriate use of *in vitro* methodology can be a powerful tool for assessing skin absorption. Furthermore, by using metabolically viable skin preparations as the diffusional membrane, the skin is no longer viewed as merely a diffusional membrane but an organ capable of many metabolic and toxicologic interactions. Consequently, well designed and executed *in vitro* experiments will produce information that has acceptable *in vivo* relevance and contribute to our understanding of the complex processes of skin absorption and disposition of topical agents.

2. *In Vivo* Studies

It is generally recognized that the most reliable method for learning about skin absorption is to measure penetration *in vivo* using the appropriate animal model or human volunteers. In principle the *in vivo* approach is simple, but in practice it is often fraught with experimental and ethical difficulties, particularly when studies are conducted in man. Typically, *in vivo* studies are performed by applying the compound of interest, in a suitable vehicle, to the surface of a defined area of skin. To protect the application site, occlusive or nonocclusive covering is often placed over the treated skin area, and absorption is then monitored by various procedures. However, the techniques used to monitor *in vivo* skin absorption often assess absorption indirectly, and frequently measurements are based on nonspecific assays; consequently, the validity of the *in vivo* determination will depend on the validity of the method used.

When a topically applied compound induces a biological response following skin absorption, the quantitation of that response may provide a basis for assessing skin absorption. Indeed such physiological or pharmacological responses have been employed as end points in assessing skin absorption *in vivo*, and perhaps the most successful example is the vasoconstrictor response to topical steroids.[22,23] However, while these pharmacodynamic endpoints may be very sensitive and selective for defined classes of compounds, it should be noted that the parameter measured is the product of both the quantity and potency of the compound under investigation, and may not necessarily reflect the extent of skin absorption, or indeed cutaneous metabolism and disposition.

Ideally skin absorption and metabolism should be assessed based on the analysis of the compound and metabolites of interest in the body following topical application, and such analysis should be performed using sensitive, selective, and specific assays. Although it has been possible in some select cases[61,62,87-89] to determine, for example, a plasma concentration-time profile of the compound following topical application, and to assess bioavailability by comparing this data with the corresponding data obtained following intravenous administration; such specific analyses in body fluids are not routinely feasible because the low absolute amount normally absorbed via the skin is often beyond the limitations of current analytical techniques. Consequently in the assessment of skin absorption *in vivo*, radiolabeled compounds are frequently used, and the extent of absorption is assessed typically by monitoring the elimination of radioactivity in excreta over a period of several days. For small laboratory animals, the absorbed radioactivity which may be retained in the animal and not eliminated in the excreta can be determined directly by analysis of the carcass following removal of the application site and appropriate solubilization.[90] However, in larger animals and in man such an approach is impractical, and a correction is required to adjust for such pharmacokinetic factors as absorption, distribution, metabolism, and excretion which may have an impact on the validity of the skin absorption assessment based on eliminated radioactivity. This correction has often been made by injecting intravenously a single dose of the radiolabeled compound and monitored radioactivity elimination in the excreta. This has been the standard approach by which the vast majority of *in vivo* skin absorption studies are conducted, and has provided invaluable information concerning percutaneous absorption in man.[91-94] However, this approach has obvious limitations and flaws, and caution must be exercised when interpreting the results of such *in vivo* studies.

The most obvious limitation is that the absorption is determined indirectly; its measurement is based on radioactivity so that the chemical nature of the penetrant is unknown. Furthermore, when measurements from intravenous dosing are applied as a correction, its validity is dependent on the underlying assumption that metabolism and disposition of the applied compound is route independent, and that the pharmacokinetic behavior of the intravenous and topical doses are similar. Unfortunately, there is little or no experimental basis for substantiating this assumption, and often the pharmacokinetic profile of the compound under investigation has not been fully characterized. Skin absorption often resembles a slow infusion, but the intravenous dose for correction is often given as a single bolus injection. Subcutaneous injection or a slow intravenous infusion may be the more appropriate delivery method for correction. Moroeover, the selection of the size of the intravenous dose is often not rationalized, and when differences in the relative amount of radioactivity excreted in the urine and feces following intravenous and topical administration are observed, these differences may be the consequence of route of administration, or they may be related to differences in extent of systemic exposure. Furthermore, when metabolites are found in the excreta following topical application it is difficult if not impossible to differentiate between skin metabolism and systemic metabolism; consequently, the significance of cutaneous metabolism in skin absorption cannot be readily established from *in vivo* investigations.

More direct approaches for monitoring skin absorption have been proposed, and measuring the rate of disappearance of the chemical at the application site is one example. However, the generally low permeability of the skin means that the rate of disappearances is often very slow, and the accuracy of the measurement will depend on analytical techniques that are capable of accurately quantifying minute differences. In reality such an approach is of limited use, and reliable results can only be obtained with chemicals that are rapidly absorbed. The main use of this technique is monitoring the loss of radioactivity from the skin surface; but it should be appreciated that measurements using high energy emitters whose transmission range may be similar to or greater than the thicknesses of the skin could result in erroneous estimates of skin absorption. Other methods, such as those based on histochemical and fluorescence techniques, are highly specialized and compound dependent,[22,23] and the recently described approach where the extent of percutaneous absorption was correlated to the reservoir function of the stratum corneum,[95] are interesting alternatives, but their general utility for estimating skin absorption have yet to be established. Nevertheless, these potential methods provide opportunities whereby the disposition of chemicals within the skin may be evaluated. As alluded to earlier for *in vitro* studies, these measurements should be considered as integral parts in an overall estimate of skin absorption.

In most toxicological and pharmacological investigations, the dose administered is precisely defined and dose response relationships are usually carefully evaluated. In percutaneous absorption studies, however, this is not always the case, and in the literature a great deal of absorption information may be of questionable validity since the dose applied was frequently not clearly defined or reported even though the extent of skin absorption is frequently reported in terms of a percentage of the dose applied. Dose application in skin absorption studies conducted *in vivo* is in general relatively straightforward. The compound of interest is prepared in an appropriate vehicle which may be liquid or semisolid, and an appropriate amount of this preparation is then applied uniformly onto the surface of the skin. Uniformity of application is important but often difficult to assess and is generally assumed without supporting evidence. Nevertheless, based on a limited number of studies it is recognized that the extent of skin absorption is greatly dependent upon the concentration of the applied dose and the surface area of exposure.[96] Increasing the concentration of the applied dose has been shown to result in a decrease in the percentage of the applied dose being absorbed, but total absorption is increased.[97] This effect may be compound specific and may depend on the dose range under investigation. Moroeover, increasing the surface area of exposure will also result in increases in the extent of absorption.[98] In defining the dose applied therefore, one must consider not only the amount of chemical applied per unit area, but also the total surface area of application and the total dose applied. The frequency of application and the duration of exposure have also been shown to influence the extent of skin absorption.[99-101] In the few times that it has been investigated, the results have shown that washing of the application site to remove the applied dose may enhance, reduce, or have no effect on absorption.[102-104] Studies on the interrelationship and influence of the various parameters pertaining to dose application in skin absorption are in their infancy. How these parameters may influence the extent of skin absorption is being explored, and it is clear that our current knowledge in this area is far from complete. In our conventional practices, *in vivo* skin absorption results are frequently normalized and reported as the percent of the applied dose absorbed; however, in assessing the validity of skin absorption measurements and dose response relationships from topical exposure, definition of the dose applied and amount absorbed requires careful consideration.

Defining the amount of the topical dose applied that is available for absorption is particularly challenging when the compound under investigation is volatile or semivolatile as in the case of solvents and insect repellents. Following topical application, these compounds will penetrate the skin and be absorbed; at the same time, they will evaporate slowly from the surface of the skin and be lost, and will not be available for percutaneous absorption. It has been demonstrated that the rate of evaporation, and consequently the relationship between evaporation and skin penetration, can influence the quantity of chemical absorbed dermally.[105-108] Therefore, since the extent of evaporation from the skin surface is a function of the dose applied, airflow, and temperature of the skin surface,[109] the extent to which these variables may be controlled or monitored can have a major impact on the results of *in vivo* skin absorption studies. Furthermore, consideration of the evaporative loss of the applied dose will be particularly important when surface disappearance or stratum corneum concentrations are employed as methods for assessing *in vivo* skin absorption.

It is evident that loss of chemicals under investigation, as in surface evaporation, or adsorption to the surface of the skin and subsequently loss by exfoliation would reduce the dose available for skin absorption. These changes may have serious implications in the assessment of skin absorption, particularly when absorption is assessed in terms of percent of applied dose, and once again raises the issue of the importance of mass balance and dose accountability determinations. As alluded to in the previous section, these studies are often not reported or conducted even though a low total recovery of the applied material may compromise the validity and interpretation of the studies. Fortunately, these issues are being addressed, and devices have been developed which provide refinements to *in vivo* methodology so as to permit accurate mass balance determinations.[110,111]

Vehicle as a modulating factor that can influence skin absorption has been discussed in great detail, particularly from a standpoint of increasing absorption in the delivery of dermatopharmaceutics,[112-114] and there is much interest in solvents such as dimethylsulfoxide and azone as vehicles because they act as penetrant enhancers,[115-117] however, their mechanism of action remains to be elucidated. Post-application loss of volatile components in the vehicles can alter the permeation characteristics of the applied chemicals; for example, if a highly volatile vehicle is used this may result in the compound under investigation being deposited as a thin film of solid onto the surface of the skin. On the other hand, a nonvolatile vehicle, such as an ointment, may be occlusive and change the diffusional properties of the stratum corneum. Both of these scenarios can greatly influence the extent of percutaneous permeation; therefore, the rationale used to justify the selection of an appropriate vehicle for dose application will have important bearing on the significance and validity of the *in vivo* observations.

It is clear, from this brief discussion, that measurements of *in vivo* skin absorption can be influenced by a variety of factors. These factors are related to the techniques and approaches used to assess the amount of the chemical of interest absorbed following topical application. The validity of these measurements as true reflection of skin absorption will depend in part on being able to distinguish between the dose applied and the dose available for absorption. Also, since the extent of skin absorption is generally derived from excretion data, experimental verification of assumptions critical to the excretion methodology will ultimately be necessary. Although this indirect approach has served as the experimental basis for much of what is known concerning factors that can influence skin absorption *in vivo*;[118] but, as it is currently conducted, the standard methodology used provides little insight into the significance of cutaneous metabolism in percutaneous absorption and subsequent disposition of topically applied agents.

3. Special Techniques

Despite the limitations, studies with traditional *in vivo* and *in vitro* methodologies, have resulted in considerable advances in our understanding of the mechanisms of percutaneous absorption. It is now generally recognized that percutaneous absorption involves both diffusional and metabolic processes. It is influenced by a variety of physiochemical and biological factors, and the overall process ultimately result in the translocation of the topically applied compounds to the superficial vasculature of the papillary dermis. At this location, the compound and/or its metabolites may remain in the dermis, or they may enter the dermal vasculature and circulate to the rest of the body. Although it has long been recognized that the microcirculation of the skin could influence this dynamic process, an accurate assessment of its role in percutaneous absorption has not been possible with the *in vivo* and *in vitro* methodologies currently in use.

Recent developments in skin flap techniques however, are providing the models which have potential utility as experimental tools for assessing the functional significance of the cutaneous vasculature in skin absorption. The skin flap procedures which have been described fall into two categories; the *in vitro* approach using the isolated perfused porcine skin flap,[119-121] and the *in vivo* approach with the rat/human skin sandwich maintained on the congenitally athymic (nude) rat.[122,123] Both of these surgical procedures produce metabolically viable and structurally intact skin preparations with clearly defined, independent and accessible vasculatures that can be used to study cutaneous absorption and metabolism. Initial results have been highly encouraging, and capillary blood flow as a relevant parameter that influence skin absorption was clearly demonstrated in studies with these models.[121-123] Potentially these skin flap approaches have many advantages over conventional methodologies;[124] but the preparation and maintenance of the skin flaps and the conduct of experiments with these models systems are relatively expensive and labor intensive. Moreover, as a true reflection of skin absorption, the validity and ultimate utility of these models have yet to be established. Nevertheless, from the view point of mechanistic investigations these models represent new research possibilities at a level of sophistication that promises to better our understanding of the dynamics of skin absorption and metabolism.

B. LIMITATIONS OF *IN VIVO* AND *IN VITRO* COMPARISONS AND EXTRAPOLATIONS

From a cursory review of the literature on percutaneous absorption it is evident that much of our current understanding on the mechanism of percutaneous absorption was derived from *in vitro* investigations. Indeed the fundamental principles governing skin absorption were defined from the results of *in vitro* experiments, and the classic treatises of Scheuplein and co-workers[9-16] have provided the theoretical basis for much of the experimental strategies currently used in the design of skin absorption studies.

The popularity and success of the traditional *in vitro* approach stem from the fact that the methodology used is relatively simple to follow. *In vitro* experiments generally afford the investigator with the ability to manipulate and control the experimental conditions, and the approach provides the unique opportunity to monitor the rate and extent of percutaneous absorption in skin tissues removed from the confounding influences of the rest of the body. Also, *in vitro* methodology provides the most practical approach currently available for studying skin absorption coupled with cutaneous metabolism, and one additional advantage of *in vitro* studies is that results can be obtained relatively quickly. The relevance of *in vitro* absorption results is larged assumed. However, the validity of *in vitro* observations as a true reflection of absorption *in vivo* is somewhat speculative, and as discussed earlier, the extent to which the *in vitro* observations will mimic *in vivo* absorption depends largely on the appropriateness of the experimental design, and the validity of the theoretical assumptions.

In the skin absorption literature there are only a few instances in which studies were designed specifically to correlate *in vivo* and *in vitro* observations.[29,79,80,125-127] This is probably because such comparative experiments involve many variables that need to be controlled or monitored, and are difficult to perform well. In general these studies, albeit with only a limited number of compounds, have demonstrated that there were good qualitative correlations between *in vivo* and *in vitro* skin permeation. In terms of the percentage of the dose applied, however, quantitative correlations were often somewhat variable. Because of the technical limitations attributed to the *in vitro* methodologies, the cause of the poor correlation was frequently and arbitrarily blamed on the inappropriate conditions of the *in vitro* studies. Some of these limitations, such as the choice of receptor fluids and the thickness of the epidermal membrane were discussed earlier. However, in the past many of the *in vitro* evaluations of skin permeation were conducted with skin preparations that were incapable of respiration, and devoid of any active biochemical processes. Therefore, the reported recovery of material diffusing through such skin preparations may only be of limited value, since it has been reported that both diffusional and metabolic processes can potentially influence the percutaneous fate of surface-applied agents.[24-28] Although the functional relevance of active processes, such as cutaneous metabolism, in skin absorption has yet to be clearly established, the use of nonviable skin preparations and inert membranes may only provide information relative to the diffusional aspects of skin absorption. The use of metabolically viable and structurally intact skin preparations, on the other hand, may generate results that are more likely to be representative of the *in vivo* situation, since the results reflect the net effects of both diffusional and metabolic processes. However, it should be appreciated that the relative importance of these processes in percutaneous absorption will depend on the properties of the compounds and the metabolic capabilities of the epidermal cells toward the compound under investigation.

As discussed earlier, *in vivo* absorption is nearly always determined indirectly; the methodology is based on excretion data, and frequently corrections based on unsubstantiated assumptions are applied to account for the fraction of the absorbed dose that was not excreted. Therefore, in reviewing the comparative literature on *in vivo* and *in vitro* skin absorption it is interesting to note that the *in vivo* results are often the standards by which the validity of the *in vitro* results, and hence, the appropriateness of the methodologies are judged. The validity of the *in vivo* results is assumed, despite the limitations of *in vivo* absorption measurements. Since both the *in vivo* and *in vitro* methodologies for assessing skin absorption are evolving and are constantly being developed and redefined, one should, therefore, exercise prudence in comparing *in vivo* results with *in vitro* observations. Meaningful comparisons can only be made when experimental parameters of the *in vitro* studies closely resemble those of the *in vivo* study or vice versa. How these parameters may affect the estimates of percutaneous absorption have been discussed in detail,[118] and some of the parameters that need to be considered include: the concentration of the applied dose, surface area of application, the total dose applied, single or multiple applications, vehicle, duration of exposure occlusion, skin site, and skin condition and treatments prior to the application of study chemicals. Sex, race and age of the subject, and the length of the *in vivo* study relative to the *in vitro* study are also important parameters for consideration.

Since man is the ultimate beneficiary of our research efforts, the general consensus is that experimental results derived from studies in man would form the most appropriate framework from which experimental findings may be extrapolated to the real world situation. However, ethical and safety concerns often limit the extent to which *in vivo* experiments may be conducted in man. *In vitro* studies with isolated human skin preparations offer an attractive alternative; but it is recognized that a major liability of human skin as a research tissue in percutaneous absorption *in vitro* is its notoriously high variability in barrier properties.[128] Frequently the source of human skin is from cadavers, and since the investigator often has little or no control over the source and characteristics of the donor skin, high variability is to be expected. Such variables as elapsed time from death to harvest of tissue, treatment and storage of the cadaver, skin care habits and health of the skin prior to death, together with the age, sex and race of the donor are potential contributors to the high variability. When skin preparations are derived from elective surgery, the preoperative procedures such as scrubbing with antimicrobial disinfectants, the surgical manipulations and the manner in which the membrane is prepared from the excised skin are also important variables that may influence the determination of *in vitro* skin absorption. Therefore, in the interpretation and extrapolation of results from *in vitro* studies with human skin, a careful assessment of the limits of the study design will be important.

Although man is the species of choice, the limitations surrounding human experimentations have led many investigators to explore animal skin as models for human skin absorption. However, species differ considerably in the structure and function of their skin, and it is unlikely that animal skin will have barrier properties that are identical to that of human skin. Therefore, a rhetorical question that is often asked is, "what is the best animal model for comparison with man?"; indeed, "which man?," since individual variations in percutaneous absorption in man have been shown to be extensive.[118] Animal skin, however, is routinely used for evaluating dermal toxicity and percutaneous absorption, and based on histological evidence and physiochemical studies it is evident that animal skin can be reasonable models that approximate human skin in percutaneous absorption. However, the debate concerning the most appropriate animal model continues, and this lack of agreement has resulted in some skepticism regarding the utility of animal studies relative to extrapolation to the real world situation.

A number of comparative studies, both *in vivo* and *in vitro*, have been conducted to identify the ideal animal model. These investigations are, of necessity, limited in scope and measurements are made only on a limited number of subjects, and moreover, the range of compounds studied are somewhat narrow. Nevertheless, based on the results reported thus far, it would appear that the ideal of animal model depends on the preference of the investigators and on the compounds under investigation. The pig, the monkey, the hairless mouse and more recently the fuzzy or hairless rat have all been described as species that showed the best potential of being predictive models of skin absorption in man.[58,129-131] However, the significance of these investigations need to be viewed within the context of the study designs and the limitations of the traditional methodologies used to measure skin absorption. In many of these investigations the results from animal studies are compared to results in man that had been previously published.[91-94] In man the site of application is frequently the ventral forearm; whereas in animals the back is often used. Moreover, in animals potentially damaging pretreatments of the skin such as shaving, clipping, or chemical depilation are frequently necessary before skin absorption experiments can be conducted. Since these are examples of variables that can influence percutaneous absorption, the reported species differences and similarities in skin absorption may reflect the net result of many competing variables, and understanding the significance of these variables would provide additional insights into the mechanism of percutaneous absorption. Although it is unlikely that any one animal will replicate the percutaneous absorption in man for all compounds, nevertheless studies in animals would increase our knowledge on the skin absorption process, and also expand the data base from which models for cutaneous absorption may be developed and improved. These models may provide the rational bases for extrapolation; however, it should be recognized that despite the advances made, the status of our current understanding of skin absorption is such that measurements in man will invariably be required for confirmation. An interesting and potentially useful development that may circumvent the limitations of human experiments and the problems of extrapolations from animals is in the use of viable human skin grafts on to the athymic nude mouse as a model for skin absorption in man.[132,133] The initial results have been encouraging, but the general utility of this approach remains to be established.

Based on our current knowledge, the important steps involved in dermal absorption have been identified as the partitioning of the compound from the delivery vehicle to the stratum corneum, transport through the stratum corneum, partitioning from the lipophilic stratum corneum into the more aqueous viable epidermis, transport across the epidermis and uptake by the cutaneous microvasculature with subsequent systemic distribution.[118] Mathematical models, of varying complexity describing skin absorption in terms of diffusion parameters have been developed.[16,34,35,134-136] From these models, mathematical expressions have been derived that relate the degree of percutaneous absorption to the physiochemical properties of the penetrants and the biophysical properties of the skin. These expressions have served as the theoretical bases for assessing percutaneous absorption; indeed, they have been invaluable as tools in the design, development and success of transdermal drug delivery systems.

Appreciation of the skin as an important drug metabolizing organ is a relatively recent phenomenon, and interest in the functional significance of cutaneous metabolism in skin absorption has led to the development of rigorous mathematical models that describe simultaneous diffusion and metabolism in biological membrane.[68-70] Unfortunately, complete evaluation of these models often requires a degree of mathematical sophistication and analytical sensitivity that cannot be matched by current experimental methodologies in skin absorption. A more pragmatic approach which permits experimental evaluation has been described; it utilizes results from both *in vivo* and *in vitro* experiments and involves the construction of linear kinetic models to describe the skin absorption processes.[137-145] In these models the generally recognized important steps in percutaneous absorption are characterized by first order rate constants. The rate constants associated with events in the skin are assumed to be proportional to important diffusional parameters, and these parameters may be estimated by *in vitro* experiments. Rate constants that are associated with systemic distribution and elimination are estimated from *in vivo* pharmacokinetic studies following intravenous administration of the penetrant. Although this approach is lacking in physiological perspectives, and the rate constants derived are subject to the limitations of the experimental methodology, nevertheless, as first approximations this approach may provide acceptable predictions of the percutaneous absorption of chemicals. The utility of this model was demonstrated recently in studies where the theoretical predictions of the plasma concentration-time profile for selected transdermally delivered drugs and chemicals compared favorably with the experimental observations.[146-149] The models have been extended to include the influence of cutaneous metabolism by adding rate constants for the formation and elimination of metabolites.[71-73] Using these extended models, computational simulations of the consequences of varying degrees of cutaneous metabolism, either by epidermal enzymes, or cutaneous microflora have been described.[150,151] Unfortunately, these simulations are only theoretical, and experimental confirmation will require a better understanding of the enzymology of cutaneous metabolism, and this will necessitate more research on the skin as a drug metabolizing organ. More recently, physiologically based perfusion-limited pharmacokinetic models have been constructed to describe percutaneous absorption, and these can be used to stimulate the transdermal delivery and cutaneous first pass metabolism of drugs.[152] The specialized skin flap methodologies may provide the ideal experimental design to assess the utility of these kinetic models in cutaneous absorption and metabolism of topically applied chemicals.

From the above discussion, it is apparent that assessing the relevance and validity of skin absorption and cutaneous metabolism studies is complex. The design, conduct, and interpretation of experimental observations require careful consideration. Furthermore, our incomplete understanding of the skin absorption process has greatly limited our ability to develop appropriate strategies for the extrapolation of experimental results to the real world situation. Nevertheless, from the multiplicity of *in vivo*, *in vitro*, and mathematical models that have been developed, we have the necessary tools with which to systematically approach the problems of assessing skin absorption. Moreoever, by considering the skin not just as a diffusional barrier, but as an organ capable of many metabolic interactions, we may gain a better understanding of the complicated process of skin absorption and disposition of topical agents.

III. DISCUSSION

It is clear that percutaneous absorption is a complex physiochemical and biological process. The skin is not just a protective envelope surrounding the body; it is a dynamic, living tissue, and as such its permeability characteristics are susceptible to constant changes. Many intrinsic and extrinsic factors are known to influence the permeability of the skin, therefore, when dealing with percutaneous absorption, the skin should not be regarded merely as an inert barrier. Instead the skin should be viewed as an organ active in many metabolic processes; indeed the significance of these cutaneous reactions in skin function are the subjects of an increasing number of investigations.

The study design determines the value of the experimental results; consequently, the interpretation of the results should not go beyond the limits of the study. Despite their limitations, traditional methodologies and study designs have provided some of the most important fundamental concepts in skin absorption. To increase our understanding of skin absorption new approaches will be required. Fortunately, techniques such as those involved with the isolation and creation of viable skin flaps with defined and accessible vasculature are being developed and validated as appropriate models. Studies with these models should further our understanding of percutaneous absorption.

Many fundamental research questions on skin absorption remain unanswered. Such questions as the potential importance of the dermal vasculature in skin absorption, the role of skin appendages and skin condition, and the influence of anatomical sites, age and disease state on skin absorption need to be addressed. The technologies to address these questions are available, and research in these areas will provide the means whereby differences in skin absorption and metabolism between species, including man, may be investigated. With an ability to directly compare the percutaneous fate of topical xenobiotics in mammalian skin under defined conditions, it should be possible to establish a basis for extrapolation and provide a predictive estimate for human skin absorption and bioavailability following topical exposure.

When skin contact with a chemical results in local effects, pathological changes in the skin may be expected to affect its barrier properties, and hence, influence the fate of surface applied chemicals. The integrity of the stratum corneum is, therefore, of primary importance. However, biochemical changes in the skin in response to topical exposure to biologically active chemicals may also influence the metabolic capabilities and metabolic status of the skin, and thereby modulate the cutaneous disposition of topically applied substances. The metabolizing activities of the skin are known to readily respond to modulation by inducers and inhibitors;[17-21,26-28,153] therefore, where there is significant cutaneous metabolism such modulation could have important implications on the outcome of cutaneous absorption and disposition of topically applied chemicals.[154,155]

From the perspective of absorption, the skin is a portal of entry for a variety of topically applied chemicals, is a drug metabolizing organ, and a target organ for local toxicity. Thus, knowledge of the processes involved in the translocation of chemicals through the skin into systemic circulation, coupled with the response of the skin to such chemicals are important aspects of skin pharmacology and toxicology. Research in this area is in its infancy and offers many opportunities. Mechanistic and functional approaches to skin absorption need to be developed. Such development will ultimately result in a better understanding of the interplay between penetration and cutaneous metabolism, and consequently, their relevance in skin toxicity and availability of topical agents. Indeed, to what extent this interrelationship may be controlled and modified remains to be established. It is anticipated that future research will increase our knowledge of skin absorption and exploitation of such knowledge would greatly facilitate the continual development of new strategies in reducing skin absorption of hazardous industrial chemicals. Also, it would provide the basis for improving topical therapy and the transdermal delivery of drugs and prodrugs.

REFERENCES

1. Standards advisory committee on cutaneous hazards findings, *Job Safety Health Rep.*, 9, 4, 1979.
2. National Institute for Occupational Safety and Health Notices 4FR 7004. *Chem. Reg. Reporter*, 3, 1666, 1980.
3. **Blank, H. I.**, Penetration of low molecular weight alcohols into skin. I. Effect of concentration of alcohol and type of vehicle, *J. Invest. Dermatol.*, 43, 415, 1964.
4. **Blank, H. I.**, Cutaneous barriers, *J. Invest. Dermatol.*, 45, 249, 1965.
5. **Blank, H. I. and Scheuplein, R. J.**, The epidermal barrier, in *Progress in the Biological Sciences in Relation to Dermatology*, Vol. 2, Rook, A. J. and Champion, R. H., Eds., Cambridge University, Cambridge, 1964, 245.
6. **Blank, H. I. and Scheuplein, R. J.**, Transport into and within the skin, *Br. J. Dermatol.*, Suppl. 4, 81, 4, 1969.
7. **Marzulli, F. N.**, Barriers to skin penetration, *J. Invest. Dermatol.*, 39, 387, 1962.
8. **Marzulli, F. N. and Tregear, R. T.**, Identification of the barrier layers in the skin, *J. Physiol.*, 157, 52P, 1961.
9. **Scheuplein, R. J.**, Mechanism of percutaneous absorption. I. Route of penetration and influence of solubility, *J. Invest. Dermatol.*, 45, 334, 1965.
10. **Scheuplein, R. J.**, Mechanism of percutaneous absorption. II. Transient diffusion and the relative importance of various routes of skin penetration, *J. Invest. Dermatol.*, 48, 79, 1967.
11. **Scheuplein, R. J.**, Permeability of the skin: a review of major concepts and some new developments, *J. Invest. Dermatol.*, 67, 672, 1976.
12. **Scheuplein, R. J. and Blank, H. I.**, Permeability of the skin, *Physiol. Rev.*, 51, 702, 1971.
13. **Scheuplein, R. J. and Blank, H. I.**, Mechanism of percutaneous absorption. IV. Penetration of nonelectrolytes (alcohols) from aqueous liquids and from pure liquids, *J. Invest. Dermatol.*, 60, 286, 1973.
14. **Scheuplein, R. J. and Ross, L. W.**, Mechanism of percutaneous absorption. V. Percutaneous absorption of solvent deposited steroids, *J. Invest. Dermatol.*, 62, 353, 1974.
15. **Scheuplein, R. J., Blank, H. I., Brauner, G. J., and MacFarlane, D. J.**, Percutaneous absorption of steroids, *J. Invest. Dermatol.*, 52, 63, 1969.
16. **Scheuplein, R. J. and Bronaugh, R. L.**, Percutaneous absorption, in *Biochemistry and Physiology of the Skin*, Vol. 2, Goldsmith, L. A., Ed., Oxford University Press, New York, 1983, 1255.
17. **Pannatier, A., Jenner, P., Testa, B., and Elter, J. C.**, The skin as a drug metabolism organ, *Drug Metab. Rev.*, 8, 319, 1878.
18. **Noonan, P. K. and Wester, R. C.**, Cutaneous biotransformation and some pharmacological and toxicological implications, in *Dermatotoxicology*, 2nd ed., Marzulli, F. N. and Maibach, H. I., Eds., Hemisphere, New York, 1983, 71.
19. **Bickers, D. R.**, Drug, carcinogen and steroid hormone metabolism in the skin, in *Biochemistry and Physiology of the Skin*, Vol. 2, Goldsmith, L. A., Ed., Oxford University Press, New York, 1983, 1169.
20. **Wester, R. C. and Maibach, H. I.**, Dermatopharmacokinetics: a dead membrane or a complex multifunctional viable process?, in *Progress in Drug Metabolism*, Vol. 9, Bridges, J. W. and Chasseaud, L. F., Eds., Taylor and Francis, London, 1986, 95.

21. **Martin, R. J., Denyer, S. P., and Hadgraft, J.,** Skin metabolism of topically applied compounds, *Int. J. Pharmaceut.,* 39, 23, 1987.
22. **Schaefer, H., Zesch, A., and Stuttgen, G.,** *Skin Permeability,* Springer-Verlag, New York, 1982.
23. **Barry, B. W.,** *Dermatological Formulation: Percutaneous Absorption,* Marcel Dekker, New York, 1983.
24. **Smith, L. H. and Holland, J. M.,** Interaction between benzo(a)pyrene and mouse skin in organ culture, *Toxicology,* 21, 47, 1981.
25. **Holland, J. M., Kao, J., and Whitaker, M. S.,** A multisample apparatus for the kinetic evaluation of skin penetration in vitro: influence of viability and metabolic status of the skin, *Toxicol. Appl. Pharmacol.,* 68, 206, 1984.
26. **Kao, J., Hall, J., Shugart, L. R., and Holland, J. M.,** An in vitro approach to studying the cutaneous metabolism and disposition of topically applied xenobiotics, *Toxicol. Appl. Pharmacol.,* 75, 289, 1984.
27. **Kao, J., Patterson, F. K., and Hall, J.,** Skin penetration and metabolism of topically applied chemicals in six mammalian species, including man: an in vitro study with benzo(a)pyrene and testosterone, *Toxicol. Appl. Pharmacol.,* 81, 502, 1985.
28. **Kao, J. and Hall, J.,** Skin absorption and cutaneous first pass metabolism of topical steroids: in vitro study with mouse skin in organ culture, *J. Pharm. Exp. Ther.,* 241, 482, 1987.
29. **Franz, T. J.,** The finite dose technique as a valid in vitro model for studying percutaneous absorption in man, *Curr. Probl. Dermatol.,* 7, 58, 1978.
30. **Cooper, E. R.,** Increased skin permeability for lipophilic molecules, *J. Pharm. Sci.,* 73, 1153, 1984.
31. **Akhter, S. A., Bennett, S. L., Walker, I. L., and Barry, B. W.,** An automated diffusion apparatus for studying skin penetration, *Int. J. Pharmaceut.,* 21, 17, 1984.
32. **Bronaugh, R. L. and Stewart, R. F.,** Method for percutaneous absorption. IV. The flow through diffusion cell, *J. Pharm. Sci.,* 74, 64, 1985.
33. **Tregear, R. T.,** *Physical Function of the Skin,* Academic Press, New York, 1966.
34. **Dugard, P. H.,** Skin permeability theory in relation to measurements of percutaneous absorption in toxicology, *Dermatotoxicology,* 2nd ed., Marzulli, F. N. and Maibach, H. I., Eds., Hemisphere, New York, 1981, 91.
35. **Flynn, G. L.,** Mechanism of percutaneous absorption from physiochemical evidence, in *Percutaneous Absorption,* Bronaugh, R. L. and Maibach, H. I., Eds., Marcel Dekker, New York, 1985, 17.
36. **Cooper, E.,** Vehicle effects on skin penetration, in *Percutaneous Absorption,* Bronaugh, R. L. and Maibach, H. I., Eds., Marcel Dekker, New York, 1985, 525.
37. **Schalla, W. and Schaefer, H.,** Localization of compounds in different skin layers and its use as an indicator of percutaneous absorption, in *Percutaneous Absorption,* Bronaugh, R. L. and Maibach, H. I., Eds., Marcel Dekker, New York, 1985, 281.
38. **Middleton, M. C. and Hasmall, R.,** A rapid method for preparing epidermal slices of reproducible thickness from excised rat skin, *J. Invest. Dermatol.,* 68, 108, 1977.
39. **Bronaugh, R. L. and Maibach, H. I.,** In vitro percutaneous absorption, in *Dermatotoxicology,* 2nd ed., Marzulli, F. N. and Maibach, H. I., Eds., Hemisphere, New York, 1981, 117.
40. **Hawkin, G. S. and Reifenrath, W. G.,** Development of an in vitro model for determining the fate of chemicals applied to the skin, *Fund. Appl. Toxicol.,* 4, S133, 1984.
41. **Bronaugh, R. L. and Stewart, R. F.,** Method for in vitro percutaneous absorption studies. VI. preparation of the barrier layer, *J. Pharm. Sci.,* 75, 487, 1986.
42. **Marty, J. P., Guy, R. H., and Maibach, H. I.,** Percutaneous penetration as a method of delivery to muscle and other tissues, in *Percutaneous Absorption,* Bronaugh, R. L. and Maibach, H. I., Eds., Marcel Dekker, New York, 1985, 469.
43. **Mackee, G. M., Sulzberger, M. B., Herrmann, F., and Bare, R. L.,** Histological studies on percutaneous penetration with special reference to the effects of vehicles, *J. Invest. Dermatol.,* 6, 43, 1945.
44. **Shelly, W. B. and Melton, F. M.,** Factors accelerating the penetration of histimine through normal intact human skin, *J. Invest. Dermatol.,* 13, 61, 1949.
45. **Tregear, R. T.,** Relative penetrability of hair follicles and epidermis, *J. Physiol. (London),* 156, 307, 1961.
46. **Rutherford, T. and Black, J. G.,** The use of autoradiography to study the localization of germicides in the skin, *Br. J. Dermatol.,* 81, suppl. 4, 75, 1969.
47. **Fredriksson, T.,** Studies on the percutaneous absorption of parathion and paraxon. II. Distribution of ^{32}P-labelled parathion within the skin, *Acta Derm. Venereol.,* 41, 344, 1961.
48. **Van Kooten, W. J. and Mali, J. W. H.,** The significance of sweat ducts in permeation experiments on isolated cadaverous human skin, *Dermatologia,* 132, 141, 1966.
49. **Wahlberg, J. E.,** Transepidermal or transfollicular absorption, In vitro studies on hairy and non-hairy guinea pig skin with sodium (^{22}Na) and mercuric (^{203}Hg) chlorides, *Acta Derm. Venereol.,* 48, 336, 1968.
50. **Wallace, S. M. and Barnett, G.,** Pharmacokinetic analysis of percutaneous absorption: evidence of parallel penetration pathways for methothrexate, *J. Pharmacokinet. Biopharm.,* 6, 315, 1978.

51. **Foreman, M. I., Picton, W., Lukowiecki, G. A., and Clark, C.,** The effect of topical coal tar treatment on unstimulated hairless hamster skin, *Br. J. Dermatol.*, 100, 707, 1979.
52. **Holland, J. M., Whitaker, M. S., and Wesley, J. W.,** Correlation of fluorescence intensity and carcinogenic potency of synthetic and natural petroleums in mouse skin, *Am. Ind. Hyg. Assoc.*, 40, 496, 1979.
53. **Kao, J., Hall, J., and Helman, G.,** In vitro percutaneous absorption in mouse skin: influence of skin appendages, *Toxicol. Appl. Pharmacol.*, 94, 93, 1988.
54. **Burch, G. E. and Winsor, T.,** Rate of insensible perspiration (diffusion of water) locally through living and through dead human skin, *Arch. Int. Med.*, 74, 437, 1944.
55. **Astley, J. P. and Levine, M.,** Effect of dimethyl sulfoxide on permeability of human skin in vitro, *J. Pharm. Sci.*, 65, 210, 1976.
56. **Harrison, S. M., Barry, B. W., and Dugard, P. H.,** Effect of freezing on human skin permeability, *J. Pharm. Pharmacol.*, 36, 261, 1984.
57. **Swarbrick, J., Lee, G., and Brom, J.,** Drug permeation through human skin I: effect of storage conditions of skin, *J. Invest. Dermatol.*, 78, 63, 1982.
58. **Kemppainen, B. W., Riley, R. T., Pace, J. G., and Hoerr, F. J.,** The effect of storage conditions and concentration of applied dose of [^{3}H]T$_2$ toxin penetration through excised human and monkey skin, *Food Chem. Toxicol.*, 24, 211, 1986.
59. **Hawkins, G. S. and Reifenrath, W. G.,** The influence of skin source, penetration cell fluid and partitation coefficient on in vitro skin penetration, *J. Pharm. Sci.*, 75, 378, 1986.
60. **Skelly, J. P., Shah, V. P., Maibach, H. I., Guy, R. H., Wester, R. C., Flynn, G., and Yacobi, A.,** FDA and AAPS report of the workshop on principles and practices of in vitro percutaneous absorption studies: relevance to bioavailability and bioequivalence, *Pharmaceut. Res.*, 4, 265, 1987.
61. **Nacht, S., Yeung, Y., Beasly, J. N., Anjo, M. D., and Maibach, H. I.,** Benzoylperoxide: percutaneous penetration and metabolic disposition, *Am. Acad. Dermatol.*, 4, 31, 1981.
62. **Wester, R. C., Noonan, P. K., Smeach, S., and Kosobud, L.,** Pharmacokinetics and bioavailability of intravenous and topical nitroglycrein in the rhesus monkey: estimate of cutaneous first pass metabolism, *J. Pharm. Sci.*, 72, 745, 1983.
63. **Kemppaninen, B. W., Riley, R. T., Pace, J. G., Hoerr, F. J., and Joyave, J.,** Evaluation of monkey skin as a model for in vitro percutaneous absorption and metabolism of [^{3}H]T$_2$ toxin in human skin *Fund. Appl. Toxicol.*, 7, 367, 1986.
64. **Grahm, M. J., Williams, F. M., and Rawlins, M. D.,** Metabolism of aldrin by rat skin in vivo. IUPHAR 9th International Congress of pharmacology, London, 1984, Abs. 138P.
65. **Grahm, M. J., Williams, F. M., Rettie, A. E., and Rawlins, M. D.,** Aldrin metabolism in skin: in vitro and in vivo studies, in *Pharmacology and the Skin*, Vol. 1, Shroot, B. and Schaefer, H., Eds., Karger, Basel 1987, 252.
66. **Hsia, S. L.,** Metabolism of steroids in human skin, in *Percutaneous Absorption of Steroids*, Mauvais-Jarvis, P., Vickers, C. F. H., and Wepierre, J., Eds., Academic Press, New York, 1980, 81.
67. **Santus, G. C., Watari, N., Hinz, R. S., Benet, L. Z., and Guy, R. H.,** Cutaneous metabolism of transdermally delivered nitroglycerin in vitro, in *Pharmacology and the Skin*, Vol. 1, Shroot, B. and Schaefer, H., Eds., Karger, Basel, 1987, 240.
68. **Ando, H. Y., Ho, N. F. H., and Higuchi, W. I.,** Skin as an active metabolizing barrier. I: Theoretical analysis of topical bioavailability, *J. Pharm. Sci.*, 66, 1525, 1977.
69. **Fox, J. L., Yu, C. D., Higuchi, W. I., and Ho, N. F. H.,** General physical model for simultaneous diffusion and metabolism in biological membranes. The computational approach for the steady state case, *Int. J. Pharm.*, 2, 41, 1979.
70. **Yu, C. D., Fox, J. L., Ho, N. F. H., and Higuchi, W. I.,** Physical model evaluation of topical prodrug delivery-simultaneous transport and bioconversion of vidaribine-5'-valerate. I. Physical model development, *J. Pharm. Sci.*, 68, 1341, 1979.
71. **Hadgraft, J.,** Theoretical aspects of metabolism in the epidermis, *Int. J. Pharmaceut.*, 4, 229, 1980.
72. **Guy, R. H. and Hadgraft, J.,** Percutaneous metabolism with saturable enzyme kinetics, *Int. J. Pharmaceut.*, 11, 187, 1982.
73. **Guy, R. H. and Hadgraft, J.,** Pharmacokinetics of percutaneous absorption with concurrent metabolism, *Int. J. Pharmaceut.*, 20, 43, 1984.
74. **Kao, J., Hall, J., and Holland, J. M.,** Quantitation of cutaneous toxicity, an in vitro approach using skin in organ culture, *Toxicol. Appl. Pharmacol.*, 68, 206, 1983.
75. **May, S. R. and DeClement, F. A.,** Development of radiometric metabolic viability testing method for human and porcine skin, *Cryobiology*, 19, 362, 1982.
76. **May, S. R. and DeClement, F. A.,** Skin banking methodology: an evaluation of package format, cooling and warming rate, and storage efficiency, *Cryobiology*, 17, 33, 1980.
77. **May, S. R. and Wainwright, J. F.,** Integrated study of the structural and metabolic degeneration of skin during 4°C storage in nutrient medium, *Cryobiology*, 22, 18, 1985.

78. **Helman, R. G., Hall, J. W., and Kao, J.,** Acute dermal toxicity: in vivo and in vitro comparison in mice, *Fund. Appl. Toxicol.,* 7, 94, 1986.
79. **Franz, T.,** Percutaneous absorption: on the relevance of in vitro data, *J. Invest. Dermatol.,* 64, 190, 1975.
80. **Bronaugh, R. L.,** Determination of percutaneous absorption by in vitro techniques, in *Percutaneous Absorption,* Bronaugh, R. L. and Maibach, H. I., Eds., Marcel Dekker, New York, 1985, 267.
81. **Brown, D. W. C. and Ulsamer, A. G.,** Percutaneous penetration of hexachlorophene as related to receptor solution, *Food Cosmet. Toxicol.,* 13, 81, 1985.
82. **Riley, R. T. and Kemppainen, B. W.,** Effect of serum parathion interactions on cutaneous penetration of parathion in vitro, *Food Chem. Toxicol.,* 23, 67, 1985.
83. **Bronaugh, R. L. and Stewart, R. F.,** Methods for in vitro percutaneous absorption studies. III. hydrophobic compounds, *J. Pharm. Sci.,* 73, 1255, 1984.
84. **Hoelgaard, A. and Mollgaard, B.,** Permeation of linoleic acid through skin in vitro, *J. Pharm. Pharmacol.,* 34, 610, 1982.
85. **Scott, R. C.,** Percutaneous absorption: in vivo, in vitro comparisons, in *Pharmacology and the Skin,* Vol. 1, Shroot, B. and Schaefer, H., Eds., Karger, Basel, 1987, 103.
86. **Tauber, U. and Rost, K. L.,** Esterase activity in skin including species variations in *Pharmacology and the Skin,* Vol. 1, Shroot, B. and Schaefer, H., Eds., Karger, Basel, 1987, 170.
87. **Vlasses, P. H., Riberio, G. T., Rotmensch, J. V., Bondi, A. E., Loper, A. E., Hchens, M., Dunlay, M. C., and Ferguson, R. K.,** Initial evaluation of transdermal Timolol; serum concentration and beta-blockade, *J. Cardiovasc. Pharmacol.,* 7, 245, 1985.
88. **MacGregor, T. R., Matzek, K. M., Keirns, J. J., Van Wayjen, R. G. A., Van den Ende, A., and van Tol, R. G. L.,** Pharmacokinetics of transdermally delivered clonidine, *Clin. Pharmacol. Ther.,* 38, 278, 1985.
89. **Laufer, L. R., DeFazio, J. L., Lu, J. K. H., Meldrum, R. D., Eggena, P., Sambhi, M. P., Hershman, J. M., and Judd, J. L.,** Estrogen replacement therapy by transdermal estradiol administration, *Am. J. Obstet. Gynecol.,* 146, 533, 1983.
90. **Shah, P. V. and Guthrie, F. E.,** Percutaneous penetration of three insecticides in rats: a comparison of two methods for in vivo determination, *J. Invest. Dermatol.,* 80, 291, 1983.
91. **Feldmann, R. J. and Maibach, H. I.,** Regional variation in percutaneous penetration of [^{14}C]cortisol in man, *J. Invest. Dermatol.,* 48, 181, 1967.
92. **Feldmann, R. J. and Maibach, H. I.,** Percutaneous penetration of steroids in man, *J. Invest. Dermatol.,* 52, 89, 1969.
93. **Feldmann, R. J. and Maibach, H. I.,** Absorption of some organic compounds through the skin in man, *J. Invest. Dermatol.,* 54, 399, 1970.
94. **Feldmann, R. J. and Maibach, H. I.,** Percutaneous penetration of some pesticides and herbicides in man, *Toxicol. Appl. Pharmacol.,* 28, 126, 1974.
95. **Rougier, A. and Lotte, C.,** Correlations between horny layer concentration and percutaneous absorption, in *Pharmacology and the Skin,* Vol. 1, Shroot, B. and Schaefer, H., Eds., Karger, Basel, 1987, 81.
96. **Wester, R. C. and Maibach, H. I.,** Interrelationships in the dose response of percutaneous absorption, in *Percutaneous Absorption,* Bronaugh, R. L. and Maibach, H. I., Eds., Marcel Dekker, New York, 1985, 347.
97. **Wester, R. C. and Maibach, H. I.,** Relationship of topical dose and percutaneous absorption in rhesus monkey and man, *J. Invest. Dermatol.,* 67, 518, 1976.
98. **Noonan, P. K. and Wester, R. C.,** Percutaneous absorption of nitroglycerin, *J. Pharm. Sci.,* 69, 385, 1980.
99. **Wester, R. C., Noonan, P. K., and Maibach, H. I.,** Frequency of application on percutaneous absorption of hydrocortisone, *Arch. Dermatol. Res.,* 113, 620, 1977.
100. **Wester, R. C., Bucks, D. A. W., Maibach, H. I., and Guy, R. H.,** Malathion percutaneous absorption after repeated administration to man, *Toxicol. Appl. Pharmacol.,* 68, 116, 1983.
101. **Bucks, D. A. W., Maibach, H. I., and Guy, R. H.,** Percutaneous absorption of steroids: effect of repeated application, *J. Pharm. Sci.,* 74, 1337, 1985.
102. **Wester, R. C. and Maibach, H. I.,** Advances in percutaneous absorption, in *Cutaneous Toxicity,* Drill, V. A. and Lazar, P., Eds., Raven Press, New York, 1983, 29.
103. **Wester, R. C. and Maibach, H. I.,** Dermal decontamination and percutaneous absorption, in *Percutaneous Absorption,* Bronaugh, R. L. and Maibach, H. I., Eds., Marcel Dekker, New York, 1985, 327.
104. **Reifenrath, W. G., Mershon, M. M., Brinkley, F. B., Miura, G. A., Broomfield, C. A., and Cranford, H. B.,** Evaluation of diethyl malonate as a simulant for 1,2,2-trimethylpropyl methylphosphonofluoridate (soman) in shower decontamination of the skin, *J. Pharm. Sci.,* 73, 1388, 1984.
105. **Spencer, T. S., Hill, J. A., Maibach, H. I., and Feldmann, R. J.,** evaporation of diethyl-toluamide from human skin in vivo and in vitro, *J. Invest. Dermatol.,* 72, 317, 1979.
106. **Reifenrath, W. G. and Robinson, P. B.,** In vitro skin evaporation and penetration characteristics of mosquito repellent, *J. Pharm. Sci.,* 71, 1014, 1982.

107. **Susten, A. S., Dames, B. L., and Niemeier, R. W.,** In vivo percutaneous absorption studies of volatile solvents in hairless mice. I. Description of a skin depot, *J. Appl. Toxicol.*, 6, 43, 1986.

108. **Guy, R. H. and Hadgraft, J.,** Percutaneous absorption kinetics of topically applied agents liable to surface loss, *J. Soc. Cosmet. Chem.*, 35, 103, 1984.

109. **Reifenrath, W. G. and Spencer, T. S.,** Evaporation and penetration from skin, in *Percutaneous Absorption*, Bronaugh, R. L. and Maibach, H. I., Eds., Marcel Dekker, New York, 1985, 305.

110. **Guy, R. H., Bucks, D. A. W., McMaster, J. R., Villaflor, D. A., Roskos, K. V., Hinz, R. S., and Maibach, H. I.,** Kinetics of drug absorption across human skin in vivo: development of methodology, in *Pharmacology and the Skin*, Vol. 1, Shroot, B. and Schaefer, H., Eds., Karger, Basel 1987, 70.

111. **Bucks, D. A. W., Maibach, H. I., and Guy, R. H.,** Mass balance and dose accountability in percutaneous absorption studies: Development of nonocclusive application system. *Pharmaceut. Res.*, 5, 313, 1988.

112. **Barry, B. W.,** Optimizing percutaneous absorption, in *Percutaneous Absorption*, Bronaugh, R. L. and Maibach, H. I., Eds., Marcel Dekker, New York, 1985, 489.

113. **Zatz, J. L., and Sarpotdar, P. P.,** Influence of vehicles on skin penetration in *Transdermal Delivery of Drugs*, Vol. 2, Kydonieus, A. F. and Berner, B. Eds., CRC Press, Boca Raton, FL, 1987, 85.

114. **Cooper, E. R.,** Alternation in skin permeability, in *Transdermal Controlled Systemic Medications*, Chien, Y. W., Eds., Marcel Dekker, New York, 1987, 83.

115. **Cooper, E. R. and Berner, B.,** Penetration enhancers, in *Transdermal Delivery of Drugs*, Vol. 2, Kydonieus, A. F. and Berner, B., Eds., CRC Press, Boca Raton, FL, 1987, 57.

116. **Vaidyanathan, R., Rajadhyaksha, V., Kim, B., and Anisko, J. J.,** Azone in *Transdermal Delivery of Drugs*, Vol. 2, Kydonieus, A. F. and Berner, B., Eds., CRC Press, Boca Raton, FL, 1987, 63.

117. **Gummer, C. L.,** Vehicles as penetration enhancers, in *Percutaneous Absorption*, Bronaugh, R. L. and Maibach, H. I., Eds., Marcel Dekker, New York, 1985, 561.

118. **Wester, R. C. and Maibach, H. I.,** Cutaneous pharmacokinetics: 10 steps to percutaneous absorption, *Drug Metabol. Rev.*, 14, 169, 1983.

119. **Riviere, J. E., Bowman, K. F., Monteiro-Riviere, N. A., Dix, L. P., and Carver, M. P.,** The isolated perfussed porcine skin flap (IPPSF). I. A novel in vitro model for percutaneous absorption and cutaneous toxicology studies, *Toxicol. Appl. Pharmacol.*, 7, 444, 1968.

120. **Monteiro-Riviere, N. A., Bowman, K. F., Scheidt, V. J., and Riviere, J. E.,** The isolated perfused porcine skin flap (IPPSF) II: Ultrastructural and histological characterization of epidermal viability, *In Vitro Toxicol.*, 1, 241, 1987.

121. **Carver, M. P., Williams, P. L., and Riviere, J. E.,** The isolated perfused porcine skin flap (IPPSF) III: percutaneous absorption pharmacokinetics of organophosphates, steroids, benzoic acid and caffeine, *Toxicol. Appl. Pharmacol.*, (in press 1989).

122. **Kreuger, G. G., Wojciechowski, Z. J., Burton, S. A., Gilhar, A., Huether, S. E., Leonard, L. G., Rohr, U. D., Petelenz, T. J. Higuchi, W. I., and Pershing, L. K.,** The development of a rat/human skin flap served by a defined and accessible vasculature on a congenitally athymic (nude) rat, *Fund. Appl. Toxicol.*, 5, S112, 1985.

123. **Wojciechowski, Z., Pershing, L. K., Huether, S., Leonard, L., Burton, S. A., Higuchi, W. I., and Krueger, G. G.,** An experimental skin sandwich flap on an independent vascular supply for the study of percutaneous absorption, *J. Invest. Dermatol.*, 88, 439, 1987.

124. **Pershing, L. K. and Krueger, G. G.,** New animal models for bioavailability studies, in *Pharmacology and the Skin*, Vol. 1, Shroot, B. and Schaefer, H., Eds., Karger, Basel 1987, 57.

125. **Bronaugh, R. L. and Franz, T. J.,** Vehicle effects on percutaneous absorption: in vivo and in vitro comparison with human skin, *Br. J. Dermatol.*, 115, 1, 1986.

126. **Bronaugh, R. L., Stewart, R. F., Congdon, E. R., and Giles, A. L.,** Methods for in vitro percutaneous absorption studies. I. comparison with in vivo results, *Toxicol. Appl. Pharmacol.*, 62, 474, 1982.

127. **Guzek, D. B., Kennedy, A. H., McNeill, S. C., Wakshull, E., and Potts, R. O.,** Transdermal drug transport and metabolism. I. Comparison of in vitro and in vivo results, *Pharmaceut. Res.*, (in press).

128. **Southwell, D., Barry, B. W., and Woodford, R.,** Variations in permeability of human skin within and between specimens, *Int. J. Pharmaceut.*, 18, 299, 1984.

129. **Bronaugh, R. L., Stewart, R. F., and Congdon, E. R.,** Methods for in vitro percutaneous absorption studies II: animal models for human skin, *Toxicol. Appl. Pharmacol.*, 62, 481, 1982.

130. **Wester, R. C. and Maibach, H. I.,** Animal models for transdermal delivery, in *Transdermal Delivery of Drugs*, Vol. 1, Kydonieus, A. F. and Berner, B., Eds., CRC Press, Boca Raton, FL, 1987, 61.

131. **Rougier, A., Lotte, C. and Maibach, H. I.,** The hairless rat; a relevant animal model to predict in vivo percutaneous absorption in human?, *J. Invest. Dermatol.*, 88, 577, 1987.

132. **Reifenrath, W. G., Chellquist, E. M., Shipwash, E. A., and Jederberg, W. W.,** Evaluation of animal models for prediction of skin penetration in man, *Fund. Appl. Toxicol.*, 4, S224, 1984.

133. **Reifenrath, W. G., Chellquist, E. M., Shipwash, E. A., Jederberg, W. W., and Krueger, G. G.,** Percutaneous penetration in the hairless dog, wealing pig, and grafted athymic nude mouse: evaluation of models for evaluation for predicting skin penetration in man, *Br. J. Dermatol.*, 111, Suppl 27, 123, 1984.

134. **Michaels, A. S., Chandrasekaran, S. K., and Shaw, J. E.,** Drug permeation through human skin: theory and in vitro experimental measurement, *AIChE. J.,* 21, 985, 1975.

135. **Chandrasekaran, S. K., Bayne, W., and Shaw, J. E.,** Pharmacokinetics of drug permeation through human skin, *J. Pharm. Sci.,* 67, 1370, 1978.

136. **Zatz, J. L.,** Percutaneous absorption: computer simulation using multicompartmented membrane models, in *Percutaneous Absorption,* Bronaugh, R. L. and Maibach, H. I., Eds., Marcel Dekker, New York, 1985, 165.

137. **Guy, R. H., Hadgraft, J., and Maibach, H. I.,** A pharmacokinetic model for percutaneous absorption, *Int. J. Pharm.,* 11, 119, 1982.

138. **Guy, R. H. and Hadgraft, J.,** Physiochemical interpretation of the pharmacokinetics of percutaneous absorption, *J. Pharmacokinet. Biopharm.,* 11, 189, 1983.

139. **Kubota, K. and Ishizaki, T.,** A theoretical consideration of percutaneous drug absorption, *J. Pharmacokinet. Biopharm.,* 13, 55, 1985.

140. **Guy, R. H. and Hadgraft, J.,** Transdermal drug delivery: a simplified pharmacokinetic approach, *Int. J. Pharmaceut.,* 24, 267, 1985.

141. **Guy, R. H. and Hadgraft, J.,** Prediction of drug disposition kinetics in skin and plasma following topical administration, *J. Pharm. Sci.,* 73, 883, 1984.

142. **Sato, K., Oda, T., Sugibayashi, K., and Morimoto, Y.,** Estimation of blood concentration of drug after topical application from in vitro skin permeation data. II. Approach by using diffusion model and compartment model, *Chem. Pharm. Bull.,* 36, 2624, 1988.

143. **Guy, R. H. and Hadgraft, J.,** The predication of plasma levels of drugs following transdermal application, *J. Control. Release,* 1, 177, 1985.

144. **Guy, R. H., Hadgraft, J., and Maibach, H. I.,** Percutaneous absorption: multidose pharmacokinetics, *Int. J. Pharmaceut.,* 17, 23, 1983.

145. **Guy, R. H., Hadgraft, J., and Maibach, H. I.,** Transdermal absorption kinetics: a physiochemical approach, in *Risk Determination for Agricultural Workers from Dermal Exposure to Pesticides,* ACS Symposium Series, ACS, Washington DC, 1985, 19.

146. **Guy, R. H. and Hadgraft, J.,** Pharmacokinetic interpretation of the plasma levels of clonidine following transdermal delivery, *J. Pharm. Sci.,* 74, 1016, 1985.

147. **Guy, R. H. and Hadgraft, J.,** Kinetic analysis of transdermal nitroglycerin delivery, *Pharmaceut. Res.,* 2, 206, 1985.

148. **Guy, R. H. and Hadgraft, J.,** Interpretation and predication of the kinetics of transdermal drug delivery: oestradiol, hyoscine, and timolol, *Int. J. Pharmaceut.,* 32, 159, 1986.

149. **Guy, R. H., Hadgraft, J., and Maibach, H. I.,** Percutaneous absorption in man: a kinetic approach, *Toxicol. Appl. Pharmacol.,* 78, 123, 1985.

150. **Guy, R. H., Hadgraft, J., and Bucks, D. A. W.,** Transdermal drug delivery and cutaneous metabolism, *Xenobiotica,* 17, 325, 1987.

151. **Denyer, S. P., Guy, R. H., Hadgraft, J., and Hugo, W. B.,** The microbial degradation of topically applied drugs, *Int. J. Pharmaceut.,* 26, 89, 1985.

152. **Nakashima, E., Noonan, P. K., and Benet, L. Z.,** Transdermal bioavailability and first pass skin metabolism: a preliminary evaluation with nitroglycerin, *J. Pharmacokinet. Biopharm.,* 15, 423, 1987.

153. **Finnen, M. J., Herdman, M. L., and Shuster, S.,** Induction of drug metabolising enzymes in the skin by topical steroids, *J. Steroid. Biochem.,* 20, 1169, 1984.

154. **Cheung, Y. W., Po, A. L. W., and Irwin, W. J.,** Cutaneous biotransformation as a parameter in the modulation of the activity of topical corticosteroids, *Int. J. Pharmaceut.,* 26, 175, 1985.

155. **Johansen, M., Mollgaard, B., Wotton, P. K., Larsen, C., and Hoelgaard, A.,** In vitro evaluation of dermal prodrug delivery transport and bioconversion of a series of aliphatic esters of metronidazole, *Int. J. Pharmaceut.,* 32, 199, 1986.

INDEX

A

B

C

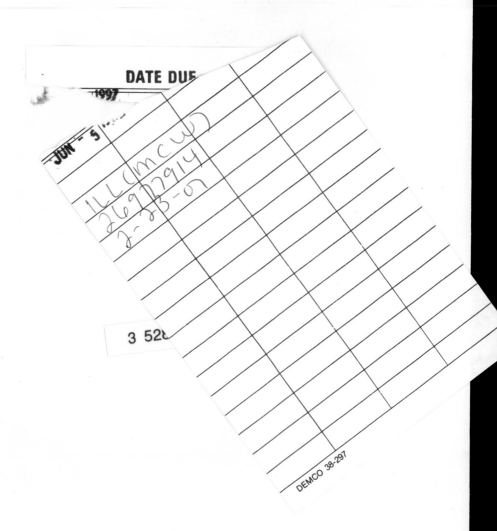